FOOD ENGINEERING LABORATORY MANUAL

FOOD ENGINEERING LABORATORY MANUAL

Gustavo V. Barbosa-Cánovas, Ph.D.
Li Ma, Ph.D.
Blas Barletta, M.S.

Washington State University
Department of Biological Systems Engineering
Pullman, Washington

CRC Press
Taylor & Francis Group
Boca Raton London New York

CRC Press is an imprint of the
Taylor & Francis Group, an **informa** business

Food Engineering Laboratory Manual

First published 1997 by Technomic Publishing Company, Inc.

Published 2019 by CRC Press
Taylor & Francis Group
6000 Broken Sound Parkway NW, Suite 300
Boca Raton, FL 33487-2742

Visit the Taylor & Francis Web site at
http://www.taylorandfrancis.com

and the CRC Press Web site at
http://www.crcpress.com

Main entry under title:
  Food Engineering Laboratory Manual

Bibliography: p.
Includes index p. 137

Library of Congress Catalog Card No. 97-60700

To our families

TABLE OF CONTENTS

The purpose of this laboratory manual is to facilitate the understanding of the most relevant unit operations in food engineering. The first chapter presents information on how to approach laboratory experiments; topics covered include safety, preparing for a laboratory exercise, effectively performing an experiment, properly documenting data, and preparation of laboratory reports. The following eleven chapters cover unit operations centered on food applications: dehydration (tray, spray, and freeze dehydration), thermal processing, friction losses in pipes, freezing, extrusion, evaporation, and physical separations. These chapters are systematically organized to include the most relevant theoretical background pertaining to each unit operation, the objectives of the laboratory exercise, materials and methods (materials, procedures, and calculations), expected results, examples, questions, and references. The experiments presented have been designed for use with generic equipment to facilitate the adoption of this manual by all institutions teaching food process engineering.

We sincerely hope this book will be a valuable addition to the food engineering literature and will promote additional interest in this fascinating field.

GUSTAVO V. BARBOSA-CÁNOVAS
LI MA
BLAS BARLETTA

ACKNOWLEDGEMENTS

The authors want to thank Washington State University (WSU) for providing an ideal framework to develop this laboratory manual, Dr. Ralph P. Cavalieri for taking the time to review it, Mr. Frank L. Younce for providing valuable assistance in performing and developing the laboratory exercises, and Ms. Sandra Jamison and Ms. Dora Rollins for their editorial comments. Our gratitude is also extended to all WSU undergraduate and graduate students that, in one way or another, contributed to the development of this manual.

ACKNOWLEDGMENTS

The author wishes to thank Washington State University [WSU] for providing the opportunity and support for this work.

CHAPTER 1

Planning Experiments

1.1 INTRODUCTION

The principles governing the engineering aspects of food processing are the same as those applied in any engineering field in that engineers are educated to analyze, synthesize, design, and operate complex systems that manipulate mass, energy, and information to transform materials and energy into useful forms, which, in this case, are food products or food ingredients. This book is designed to give food engineering and/or food science students an understanding of the engineering principles and hands-on experiences involved in the processing of food products. With a clear understanding of the engineering basic principles of food processing, it is possible to develop new food processes and modify existing ones. Because an essential component of any laboratory exercise is to receive proper laboratory orientation, follow safety guidelines, and prepare laboratory reports, this chapter deals with fundamental engineering aspects related to specific laboratory exercises, how to get ready for an experiment, and how to report it.

1.2 MASS BALANCE

The law of conservation of mass states that mass cannot be created nor destroyed, so a mass balance in any process can be written as follows:

$$Input - Output = Accumulation$$

In a continuous process at steady state, the accumulation is zero. Therefore, a simple rule that "what goes in must come out" holds. For example, in the concentration process of milk, whole milk is fed into an evaporator. Under the law of conservation of mass, the total number of pounds of material (whole milk) entering the evaporator per unit time must equal the total number of pounds of concentrated milk and evaporated moisture that leave the evaporator. When solving the mass balance, four key steps must be followed:

(1) Select a system and draw a diagram representing the process (including all pertinent information on stream rate and compositions).
(2) Select an appropriate basis for calculation.
(3) Write the mass balance relationships for the various constituents in terms of the known and unknown quantities.
(4) Solve the resulting algebraic equations for the unknown quantities.

1.2.1 Example

A milk concentrate is to be made by evaporating water from whole milk. The whole milk contains 13% total solids (*TS*), and the concentrate should contain 49% *TS*. Calculate the amount of product and the water that needs to be evaporated.

1.2.1.1 SOLUTION

- *Step 1:* Select the evaporator to be a target system, and draw a diagram representing the process as follows:

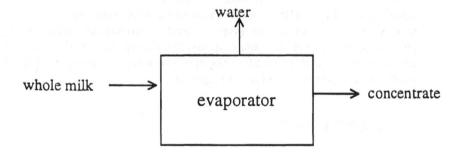

- *Step 2:* Select a basis for the calculation of 100 kg of incoming whole milk.
- *Step 3:* Write up the equations for the mass balance:
  (1) The balance for the total material:

$$100 \text{ (kg whole milk)} = W \text{ (kg water evaporated)}$$
$$+ C \text{ (kg concentrate)} \qquad (1.1)$$

(2) The balance of total solids *(TS):*

$$100(0.13) \text{ (kg } TS) = W(0) \text{ (kg TS)} + C(0.49) \text{ (kg } TS) \qquad (1.2)$$

- *Step 4:* Solve the equations. Note that Equation (1.2) has only one unknown and should therefore be solved first. The result is:

$$C = 26.5 \text{ kg}$$

Substituting the C value into Equation (1.1), we get:

$$W = 73.5 \text{ kg}$$

- *Step 5:* Appropriate answer: From every 100 kg of whole milk, we can manufacture 26.5 kg of the concentrated milk, and thus 73.5 kg of water must be evaporated.

1.3 ENERGY BALANCE

The law of conservation is also applied to energy, where its balances are employed in much the same manner as material balances:

Total energy entering system = Total energy leaving the system

Energy may appear in many forms; some of the more common are heat, work, internal energy, enthalpy, mechanical, and electrical. Heat and work are considered transitory. An additional complexity of energy balances is the need for various material property data (specific heats, latent heats of vaporization and fusion, freezing or boiling points, and process-related data such as temperatures). The following example provides an illustration of an energy balance:

1.3.1 Example

A still retort containing 1000 cans of apple sauce was sterilized at 121°C. After sterilization the cans need to be cooled down to 37°C before leaving the retort. The specific heats of the apple sauce and the can metal are 3730 and 510 J/kg K, respectively. Each can weighs 50 g and contains 450 g of apple sauce. The retort wall is made of cast iron and weighs 3000 kg. It is assumed that cooling by the surrounding air is negligible. Calculate the amount of cooling water required if it enters at 20°C and leaves at 30°C.

1.3.1.1 SOLUTION

- *Step 1:* Define the operation: The operation is cooling, and the desired result is the amount of cooling water needed.
- *Step 2:* Determine the basis of calculation: In this problem, 1000 metal cans that contain apple sauce are selected as the basis.
- *Step 3:* Write the energy balance equation: This will balance the total heat (Q) leaving the system against all the heat forms entering the system.

The heat entering the system consists of four parts: (1) heat in the cans, (2) heat in the apple sauce, (3) heat in the cooling water, and (4) heat in the retort wall. For calculation convenience, the reference temperature datum is selected at 37°C.

(1) Heat in the cans = (weight of cans)(specific heat of can)(temperature above the datum). Mathematically:

$$Q_1 = 1000 \text{ (can)} \cdot \frac{50 \text{ (g/can)}}{1000 \text{ (g/kg)}} \cdot 510 \left(\text{J/} [\text{ kg·K}] \right) \cdot (121 - 37) \text{ (K)}$$

$$= 2,142,000 \text{ J}$$

$$= 2142 \text{ kJ}$$

(2) Heat in the contents = (total weight of apple sauce)(specific heat of apple sauce)(temperature above the datum). Mathematically:

$$Q_2 = 1000 \text{ (can)} \cdot \frac{450 \text{ (g/can)}}{1000 \text{ (g/kg)}} \cdot 3730 \left(\text{J/} [\text{ kg·K}] \right) \cdot (121 - 37) \text{ (K)}$$

$$= 1.410 \times 10^8 \text{ J}$$

$$= 1.410 \times 10^5 \text{ kJ}$$

(3) Heat in the water = (weight of water)(specific heat of water)(temperature above the datum). Mathematically:

$$Q_3 = W \text{ (kg)} \cdot 4180 \left(\text{J/} [\text{ kg·K}] \right) \cdot (20 - 37) \text{ (K)}$$

$$= -71,060W \text{ (J)}$$

$$= -71.06W \text{ (kJ)}$$

(4) Heat in the retort wall:

$$Q_4 = 3000 \text{ kg} \cdot 450 \left(\text{J/}[\text{kg·K}] \right) \cdot (121 - 37) \text{ (K)}$$

$$= 113,400,000 \text{ J}$$

$$= 113,400 \text{ kJ}$$

Therefore, the total heat entering the system is:

$$Q_E = Q_1 + Q_2 + Q_3 + Q_4$$

$$= 2142 + 1.410 \times 10^5 + (-71.06W) + 113,400$$

$$= 229,542 - 71.06W \text{ (kJ)}$$

The heat leaving the system also consists of four parts: (1) heat in the cans, (2) heat in the apple sauce, (3) heat in the cooling water, and (4) heat in the retort wall.

(1') Heat in the cans = (weight of cans)(specific heat of can)(temperature above the datum). Mathematically:

$$Q_1' = 1000 \text{ (can)} \cdot \frac{50 \text{ (g/can)}}{1000 \text{ (g/kg)}} \cdot 510 \left(\text{J/}[\text{kg·K}] \right) \cdot (37 - 37) \text{ (K)}$$

$$= 0$$

(2') Heat in the can contents = (total weight of apple sauce)(specific heat of apple sauce)(temperature above the datum). Mathematically:

$$Q_2' = 1000 \text{ (can)} \cdot \frac{450 \text{ (g/can)}}{1000 \text{ (g/kg)}} \cdot 3730 \left(\text{J/}[\text{kg·K}] \right) \cdot (37 - 37) \text{ (K)}$$

$$= 0$$

(3') Heat in the water = (weight of water)(specific heat of water)(temperature above the datum). Mathematically:

$$Q_3' = W \text{ (kg)} \cdot 4180 \left(\text{J/}[\text{kg·K}] \right) \cdot (30 - 37) \text{ (K)}$$

$$= -39,260W \text{ (J)}$$

$$= -39.26W \text{ (kJ)}$$

(4′) Heat in the retort wall:

$$Q'_4 = 3000 \text{ kg} \cdot 450\big(\text{J}[\text{kg} \cdot \text{K}]\big) \cdot (37 - 37)(\text{K})$$

Therefore, the total heat leaving the system is:

$$Q_L = Q'_1 + Q'_2 + Q'_3 + Q'_4$$

$$= 0 + 0 + (-39.26\text{W}) + 0$$

$$= -39.26\text{W (kJ)}$$

According to the law of conservation:

$$Q_E = Q_L$$

$$229{,}542 - 71.06W \text{ (kJ)} = -39.26W$$

- *Step 4:* Solve the energy balance equation:

$$W = 7218 \text{ kg}$$

- *Step 5: Appropriate answer:* To cool down the apple sauce from 121°C to 37°C, 7218 kg cool water (20°C) is needed.

1.4 LABORATORY ORIENTATION

1.4.1 Laboratory Safety

   Laboratory participants need a suitable orientation to equipment and procedures to provide a safe and profitable learning experience. The safety of all persons involved is of utmost importance in laboratory exercises, which means that everyone must have knowledge of basic safety principles, work alertly, and support one another in safe laboratory practices. Before beginning any laboratory exercise, each person must be oriented to the guidelines for safe use of laboratory equipment and emergency response procedures. A checklist suggested for laboratory safety orientation is given below:

   (1) Identify locations of safety equipment:
   - fire extinguishers, fire blankets, fire alarms
   - first aid kit
   - emergency showers and eye wash stations

(2) Locate nearest exits and evacuation routes to use in case of emergency.

(3) Identify persons who are certified for administering first aid.

(4) Locate nearest telephone and emergency telephone numbers.

(5) Wear safety glasses with side shields or goggles when hazards are present:
- chemicals or liquid fuels
- flying materials from cutting or chipping processes
- nearby pressurized fluids

(6) Wear ear protection when working in noisy conditions.

(7) Do not wear gloves, long sleeves, or jewelry when near rotating machinery.

(8) Cover long hair or loose clothing when using rotating machines.

(9) Wear foot protection in the laboratory:
- no bare feet allowed
- steel-toed shoes required when handling heavy items

(10) Know characteristics of any potentially hazardous materials by checking Material Safety Data Sheets (MSDS) in the lab prior to use.

(11) Immediately report any injuries to the instructor.

(12) Know proper disposal for any chemicals used.

(13) Know potential hazards and safety practices before operating equipment by reading safety instructions for equipment to be used.

(14) Follow good housekeeping practices:
- keep work areas free from obstacles that may cause slips or falls
- keep equipment clean for optimal operation

1.4.2 Laboratory Equipment

Profitable laboratory experimentation requires equipment and instruments that perform reliably. Because laboratory equipment can be damaged by improper handling and equipment repairs and replacement are costly, users must be instructed in the proper use of equipment. Therefore, those who are unfamiliar with a given piece of equipment are expected to request assistance and/or training before attempting to use it independently.

Most equipment and instruments are supplied with operation manuals that define proper operating practices and calibration procedures. Users should become familiar with the basics of safe operation and also check with laboratory technicians to ensure the instruments have been calibrated.

1.4.3 Laboratory Teams

Because most laboratory exercises are team efforts, students must be willing and able to work collaboratively. Professional and responsible behavior is expected at all times. If a team is assigned to work together over an extended period of time, the members need to define their expectations of one another and the consequences of failing to meet the team's expectations. This will minimize misunderstandings and enhance team productivity. A sample agreement is in the Appendix.

1.5 LABORATORY REPORTS

A laboratory report is used to communicate experimental findings to others who have interest in the results of the completed tests. The format and length of the report will depend on the intended audience and desired impact, so no standard format exists for all reporting needs. However, without a report of some type, the results of the experiment and their implications will be uncertain and of value to no one. Below are two common report formats used for different target audiences.

1.5.1 Informal Report

An informal laboratory report is used to document the results of an experiment and draw useful conclusions but is not intended to become a lasting document that is widely distributed or presented to the public. It therefore includes the parts that are essential to ensuring the results are valid and useful but does not have polished textual descriptions, graphs, or persuasive summaries. An informal report should include the following:

- date and location of the work
- names of persons doing the work
- objectives or desired outcomes
- description of methods or procedures used
- data collected (data sheets, tables, printouts)
- analysis of data with appropriate statistics
- interpretation of results

1.5.2 Formal Report

A formal laboratory report is required when experimental results are to be presented to a person or group that expects to see a professional document. Because the readers of this report may have varying interest in the experimentation details, much is provided regarding how the experimentation was conducted and data interpreted. Because the format must also enable the reader to identify the essence of the experiment with minimal effort, one typically finds an abstract in the formal report, which is frequently bound to provide a professional appearance, and contains the following parts (usually in this order):

(1) *Title:* This page includes the title of the experiment, name of report writer, names of others involved in the experiment, and date on which the experiment was performed.

(2) *Abstract:* This includes a concise statement of the purpose of the experiment, major findings (e.g., property values, relationships), and recommendations.

(3) *Introduction:* This is a brief description of the experiment which should include description of the physical problem, the phenomena investigated, quantities measured, and measurement strategies used to ensure the quality of the measurements.

(4) *Objectives:* This is a statement of the purpose of the experiment and the expected results.

(5) *Materials and Methods:* This section includes apparatus, procedure, and calculation, further described below:

- *Apparatus* is a list and description of equipment or setup used in the experiment. This may include general descriptions of common items, specific data on specialized equipment, sketches of physical relationships, and/or a schematic diagram of the equipment or experimental setup.
- *Procedure* is a brief description of the experimental procedure or a list of steps used in obtaining the data. This should include types of measurements used and the number of observations made for each condition of the experiment. Estimates of expected error for each observation type may be necessary to establish the required number of observations.
- *Calculations:* If appropriate, the functional relationships of the physical quantities expressed should be stated and defined in equation form. This may include a sample calculation with the parameters, data, and calculated quantities summarized in tabular form.

(6) *Results:* The results should include data sheet(s), sample calculation(s), graph(s), other forms of experimental data obtained, values (with units) obtained for the measured quantities, and an estimate of the reliability of the results (e.g., standard deviation of observation or confidence limits). Numerical results should be presented with only the number of digits that are meaningful in light of the uncertainties found for the measurements. One should state clearly the manner in which the uncertainties in the measured results were determined, and the figures and tables must be labeled clearly and completely to avoid future uncertainty about units or which data correspond to which test conditions.

(7) *Data Analysis and Interpretation of Results:* Data analysis should include the data manipulation used to produce the form of results desired (e.g., means, conversions to other units, removal of known biases). Statistical analyses may be required to determine their derived material properties or derived equations that describe the phenomena of interest. The processed data should be tabulated or graphed to illustrate patterns or relationships, with variability (e.g., confidence intervals or standard deviation of means) defined when possible. Tables and graphs must include units, titles, and footnotes as needed to ensure the reader knows precisely what the results indicate.

Results should be interpreted relative to the initial problem statement or objectives and their practical value. This may include characterizations of phenomena observed or values of determined properties and their associated uncertainties. Both quantitative and qualitative statements are important to communicate results effectively.

(8) *Summary and Conclusions:* The discussion section gives the experimenter an opportunity to summarize and make concluding remarks that could include generalizations and opinions. It may include a discussion of the accuracy of the test procedure, possible sources of error, and a reaction to the overall laboratory experience. It may also address the quality of the measurements mentioned in the interpretation of the results section, but it should be concerned primarily with what was learned in the laboratory.

1.6 APPENDIX: SAMPLE TEAM CONTRACT

As members of an engineering team our mission is to perform our assignment with high standards that bring credit to the team and our profession and that provide results which truly reflect the phenomena we are studying. Each of us will contribute our proportional share of the total effort required, capitalizing on the strengths of each member by helping one another perform the parts that make up the whole. We agree upon the following distribution of responsibilities for this assignment:

Name	Responsibilities	Due Date

Anyone failing to complete his or her part of the assigned work by the agreed time will:

_____

_____

We, the undersigned, agree to perform our responsibilities in a professional and responsible manner that supports the success of our mutual efforts.

Name   Signature   Date

_____   _____   _____

_____   _____   _____

_____   _____   _____

_____   _____   _____

1.7 SUGGESTED REFERENCE BOOKS

1. Charm, S. E. 1978. *The Fundamentals of Food Engineering.* AVI Publ. Co. Westport, CT.
2. Toledo, R. T. 1980. *Fundamentals of Food Processing Engineering.* AVI Publ. Co. Westport, CT.
3. Heldman, D. R. and Singh, R. P. 1980. *Food Process Engineering.* 2nd ed. AVI Publ. Co. Westport, CT.

CHAPTER 2

Friction Losses Determination in a Pipe

2.1 INTRODUCTION

Transportation of fluid food is a common and essential operation in food industry applications; for example, raw milk is pumped from a storage tank to a heat exchanger for pasteurization, and tomato sauce is pumped from a storage tank to an evaporator for concentration. The energy efficiency of the system depends on the flow characteristics of the medium to be transported and the characteristics of the system's components. Greater efficiency can be achieved by minimizing friction loss during fluid food transportation. This laboratory practice will focus on the evaluation of energy loss during transportation.

In a simple transport system in which fluid food is pumped from point one to point two (Figure 2.1), an energy balance will lead to the well-known Bernoulli equation:

$$Z_1 + \frac{P_1}{\rho} + (KE)_1 + W = Z_2 + \frac{P_2}{\rho} + (KE)_2 + E_f \qquad (2.1)$$

where

Z_1, Z_2 = potential energy terms that are a function of height
$(KE)_1, (KE)_2$ = kinetic energy terms that are a function of the squared power of the fluid velocity
P_1, P_2 = static pressure terms
W = net mechanical energy transferred to or from the fluid
E_f = friction losses

This laboratory practice will focus on the evaluation of the last term on the right hand side of Equation (2.1) by studying head loss (pressure loss) within a pressurized pipeline with various fittings. The system will be under a constant head obtained from a water column of constant height. A series of pressure taps located up and downstream of pipe sections and fittings are connected to manometers, each of which directly indicates the head (expressed in inches of water) at the respective tap location in the system.

Friction losses can be predicted if enough information about the system is available. The Moody diagram (Figure 2.2) is commonly used for this purpose, but it requires the Reynolds number and information about the roughness factor of the wall of the pipe (normally supplied by the manufacturer). The friction factor (f) is obtained from the diagram, and along with the flow rate, is used to compute the friction losses (E_f). In this lab the student is expected to compare the experimental data with predicted values. There are also standard formulas and tables (see References [1] and [2]) to evaluate the losses in fittings that can be compared against experimental data. The fluid used to study friction losses in this lab is water, which is a Newtonian fluid (typical non-Newtonian fluids are tomato sauce or applesauce).

2.2 OBJECTIVES

(1) To demonstrate the fundamentals of head loss in pipe systems

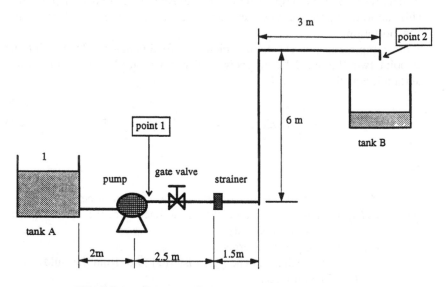

FIGURE 2.1. Schematic illustration of a fluid transport system.

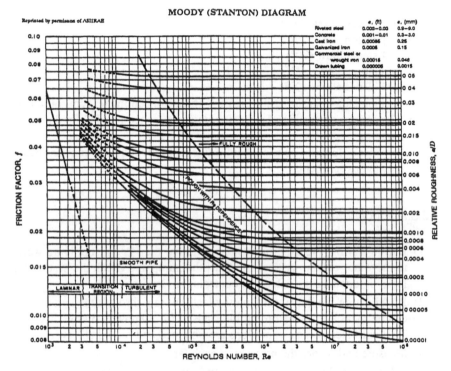

FIGURE 2.2. Moody diagram.

(2) To measure the friction loss in fittings such as elbows, valves, reductions, and expansions

(3) To measure the influence of the flow rate on the friction losses

2.3 MATERIALS AND METHODS

2.3.1 Materials

The experimental equipment consists of an arrangement of pipe pieces connected to each other using several types of fittings. The following are the units that should be installed:

- garden hose
- 10 ft of 1/2″ steel pipe
- 10 ft of 3/4″ steel pipe
- 10 ft of 1/2″ PVC pipe

- 1/2" steel elbow (90°)
- expansion 1/2" * 3/4"
- 3/4 steel elbow (90°)
- reduction 3/4" * 1/2"
- 1/2" PVC elbow (90°)
- manometers measuring the pressure (head) in centimeter (cm) of water
- flow meter consisting of a sensing device attached to a PC for data logging
- a water column of constant height to supply the required constant head for the experiments

2.3.2 Procedures

(1) Before starting the experiments, be sure that the system supplying water to the water column is working. Start the pump and check the system for leaks by opening the valve that controls the flow rate.

(2) By adjusting the valve, set a particular flow rate through the system that can be measured by recording the time to fill a known volume of water. Wait until steady-state conditions are obtained and record all raw data.

(3) Measure and record the pressure at each tap.

(4) Select a new flow rate by adjusting the control valve. Measure and record the new flow rate and pressure at each tap, repeating this step for three different flow rates.

(5) Draw a sketch of the pipe system showing the size of the tubes, location of the pressure taps, lengths of pipe, and fittings. Label your sketch with the name of each component, the material from which it is made, and its size.

2.3.3 Calculations

The energy losses of both Newtonian and non-Newtonian fluids due to friction involve straight pipes, valves, and fittings expressed as:

$$E_f = \frac{2fv^2L}{g_cD} + \sum \frac{k_fv^2}{2g_c}$$

where

D = diameter of pipe (m or ft)
L = length of pipe (m or ft)

g_c = proportionality factor, equal to 32.174 ft $lb_m/(lb_f sec^2)$ (or 1.0 in SI units)
v = fluid velocity (m/sec or ft/sec)
f = Fanning friction factor, dimensionless
k_f = friction loss coefficient, dimensionless

Note that k_f is unique for any particular valve or fitting and that different values of v, k_f, and f may be required when the pumping system's pipes have different diameters. Also, losses due to specific equipment (such as heat exchangers) must be added to E_f (see Reference [3]).

Estimating the Fanning friction factor (f) is key to the calculation of energy losses. The following is a list of the detailed procedures needed to estimate the Fanning friction factor for Newtonian and non-Newtonian fluids, respectively.

2.3.3.1 CALCULATING THE FANNING FRICTION FACTOR FOR NEWTONIAN FLUIDS

The friction factor for streamline flow (i.e., Reynolds number < 2100) can be obtained from the following relationship:

$$f = \frac{16}{Re} \tag{2.3}$$

where f is the friction factor and the dimensionless Reynolds number (Re) is defined as:

$$Re = \frac{\rho v D}{\mu} \tag{2.4}$$

where

ρ = density of the fluid
μ = viscosity of the fluid
v = average velocity of fluid in the pipe
D = pipe diameter

The average velocity can be obtained from the flow rate:

$$v = \frac{Q}{A} \tag{2.5}$$

where Q is the flow rate and A is the cross-sectional area of the pipe.

For turbulent flow the friction factor can be obtained from the Moody diagram (see Figure 2.2). The parameters required for this chart are the Reynolds number and relative roughness of the pipe, supplied by the pipe manufacturer and/or available in Reference [2].

2.3.3.2 CALCULATIONS OF THE FRICTION FACTOR FOR NON-NEWTONIAN FLUIDS

Non-Newtonian foods rarely flow under turbulent conditions, but under laminar flow conditions, the friction factor (f) may be found by the following equation:

$$f = \frac{16}{\psi(Re)_G} \tag{2.6}$$

where $(Re)_G$, the generalized Reynolds number, is:

$$(Re)_G = \frac{D^n(v^{2-n})\rho}{8^{n-1}K}\left(\frac{4n}{1+3n}\right)^n \tag{2.7}$$

$$\psi = (1+3n)^n(1-\xi_0)^{1+n}\left[\frac{(1-\xi_0)^2}{(1+3n)} + \frac{2\xi_0(1-\xi_0)}{(1+2n)} + \frac{\xi_0^2}{(1+n)}\right]^n \tag{2.8}$$

and

$$\xi_0 = \frac{2\tau_0}{f\rho v^2} = \frac{\tau_0}{\tau_w} \tag{2.9}$$

where

τ_0 = yield stress (Pa or lb_f/ft^2)
τ_w = shear stress at pipe wall (Pa or lb_f/ft^2)

ξ_0 can also be given as an implicit function of $(Re)_G$ and the generalized Hedstrom number (He):

$$(Re)_G = 2He\left(\frac{n}{1+3n}\right)^2\left(\frac{\psi}{\xi_0}\right)^{(2/n)-1} \tag{2.10}$$

$$He = \frac{D^2\rho}{K}\left(\frac{\tau_0}{K}\right)^{(2/n)-1} \tag{2.11}$$

To calculate the friction factor for Herschel-Bulkley fluids, ξ_0 is estimated through iteration of Equation (2.10) using Equations (2.7), (2.8), and (2.11). The friction factor (f) can be calculated using Equations (2.6), (2.7), and (2.8).

For power law fluids, $\xi_0 = 0$ and $\psi = 1$; hence, f can be computed directly from Equations (2.6) and (2.7). Friction factor calculations can be accomplished on a hand calculator or small computer.

Friction losses in valves and fittings can be determined by evaluating the summation term in Equation (2.6). Each valve and fitting will have a particular friction loss coefficient (k_f) associated with it. These values are readily available for Newtonian fluids (particularly water) in turbulent flow but are difficult to find for non-Newtonian fluids in laminar flow. Available data [4] indicate that friction loss coefficients increase significantly with decreasing values of the generalized Reynolds number, and this problem must be considered when pumping over short distances where friction losses in valves and fittings make a significant contribution to the overall pressure drop. More experimental data are needed to adequately deal with this problem, but common practice is to use values of k_f measured for Newtonian fluids in turbulent flow.

2.4 EXPECTED RESULTS

The following raw data and results are expected to be submitted:

(1) A complete data sheet with the raw data collected during the experiments.
(2) A graph with separate curves for each flow rate showing head in inches on the y-axis and distance from the water source in feet on the x-axis. (Note that the curve for each flow rate will not be smooth but have jumps of head at the fittings.)
(3) The theoretical friction loss for each piece of pipe and fitting, and the source from which you obtained the parameters for any calculation (include a photocopy if possible). Compare the theoretical results with the values obtained in the lab, and explain any differences between them.

2.4.1 Example

The transport system illustrated in Figure 2.1 is being used to pump applesauce at 40°C through a 1.37-inch (I.D.) smooth pipe from storage tank A to storage tank B at a rate of 25 gal/min. The strainer pressure drop is 100 kPa. The flow properties of applesauce are:

- yield stress: 157 Pa
- flow behavior index: 0.45
- consistency index: 5.20 Pa sec
- density: 1250 kg/m^3

Estimate the friction loss.

2.4.1.1 SOLUTION

Friction losses are found in the straight pipes, three long radius elbows, one open gate valve, and the strainers. The friction loss coefficients for the elbows and gate valves are 0.45 and 9.0, respectively.

$$\text{Total pipe length: } L = 2.5 + 1.5 + 6 + 3 = 13 \text{ m}$$

$$\text{Inner diameter of the pipe: } D = 1.37 \text{ in.}\left(\frac{0.0254 \text{ m}}{\text{in.}}\right) = 0.0348 \text{ m}$$

$$\text{Flow rate: } Q = 25 \text{ gal/min}\left(\frac{3.785 \times 10^{-3} \text{ m}^3}{\text{gal}} \cdot \frac{\text{min}}{60 \text{ sec}}\right)$$

$$= 1.577 \times 10^{-3} \text{ m}^3/\text{sec}$$

$$\text{Mass average velocity: } v = \frac{Q}{A} = \frac{1.577 \times 10^{-3} \text{ m}^3/\text{sec}}{\left(\dfrac{\pi}{4} \cdot 0.0348 \text{ m}\right)^2} = 1.66 \text{ m/sec}$$

From Equation (2.7):

$$(Re)_G = \frac{D^n\left(v^{2-n}\right)\rho}{8^{n-1} K}\left(\frac{4n}{1+3n}\right)^n$$

$$= \frac{(0.0348 \text{ m})^{0.45}\left(1.66^{2-0.45}\right) 1250 \text{ (kg/m}^3)}{8^{0.45-1}(5.20 \text{ Pa} \cdot \text{sec})}\left(\frac{4(0.45)}{1+3(0.45)}\right)^{0.45}$$

$$= 365$$

From Equation (2.9):

$$\xi_0 = \frac{2\tau_0}{f\rho v^2} = \frac{\tau_0}{\tau_w}$$

$$\xi_0 = \frac{2\tau_0}{f\rho v^2} = \frac{2(5.20 \text{ Pa} \cdot \text{sec})}{f(1250 \text{ kg/m}^3)(1.66 \text{ m/sec})^2}$$

$$= \frac{1}{331.20 f} \tag{2.12}$$

Substitute this result into Equation (2.8):

$$\psi = (1+3n)^n (1-\xi_0)^{1+n} \left[\frac{(1-\xi_0)^2}{(1+3n)} + \frac{2\xi_0(1-\xi_0)}{(1+2n)} + \frac{\xi_0^2}{(1+n)} \right]^n$$

$$= (1+3(0.45))^{0\,45} \left(1 - \frac{1}{331.20f}\right)^{1+0\,45}$$

$$\left[\frac{\left(1 - \frac{1}{331.20f}\right)^2}{(1+3(0.45))} + \frac{\frac{2}{331.20f}\left(1 - \frac{1}{331.20f}\right)}{(1+2(0.45))} + \frac{\left(\frac{1}{331.20f}\right)^2}{(1+(0.45))} \right]^{0\,45} \tag{2.13}$$

From Equation (2.11):

$$He = \frac{D^2\rho}{K}\left(\frac{\tau_0}{K}\right)^{(2/n)-1}$$

$$= \frac{(0.0348)\,\text{m}^2\,1250\,(\text{kg/m}^3)}{5.20\,(\text{Pa}\cdot\text{sec}^n)}\left(\frac{157\text{Pa}\cdot\text{sec}}{5.20(\text{Pa}\cdot\text{sec}^n)}\right)^{(2/0\,45)-1}$$

$$= 36,427$$

Substitute $He = 36,427$ and $(Re)_G = 365$ into Equation (2.10):

$$(Re)_G = 2He\left(\frac{n}{1+3n}\right)^2\left(\frac{\psi}{\xi_0}\right)^{(2/n)-1}$$

$$365 = 2(36,427)\left(\frac{0.45}{1+3(0.45)}\right)^2\left(\frac{\psi}{\xi_0}\right)^{(2/0\,45)-1}$$

$$\psi = 0.5611\xi_0 \tag{2.14}$$

By combining Equations (2.13) and (2.14), ξ_0 and ψ can be estimated through iteration:

$$\xi_0 = 0.508$$

$$\psi = 0.285$$

From Equation (2.6), the friction factor can be calculated:

$$f = \frac{16}{\psi\,(Re)_G} = \frac{16}{0.285\,(365)} = 0.154$$

Using Equation (2.2), the energy loss is:

$$E_f = \frac{2fv^2 L}{g_c D} + \sum \frac{k_f v^2}{2g_c}$$

$$= \frac{2(0.154)(1.66 \text{ m/sec}^2)^2(13 \text{ m})}{(1)(0.0348 \text{ m})} + \frac{3(0.45)(1.66 \text{ m/sec})^2}{2}$$

$$+ \frac{9(1.66 \text{ m/sec}^2)^2}{2} + \frac{100,000}{1250}$$

$$= 411 \text{ J/kg}$$

2.5 QUESTIONS

(1) How does the flow rate affect head loss?
(2) Which pipe material had the lowest head loss at a particular flow rate and why?
(3) Why do fittings have higher head loss than straight lengths of pipe?
(4) What effect does pipe diameter have on head loss?

2.6 REFERENCES

1. Heldman, D.R. 1981. *Food Process Engineering.* 2nd ed. AVI. Westport, CT.
2. Perry, R.H. and D. Green. 1991. *Perry's Chemical Engineer's Handbook.* 6th ed. McGraw-Hill, New York.
3. Steffe, J.F. and Morgan, R.G. 1986. Pipeline design and pump selection for non-Newtonian fluid foods. *Food Technology* 40 (12):78–85.
4. Steffe, J.F., Mohamed, I.O., and Ford, E.W. 1984. Pressure loss in valves and fittings for pseudoplastic fluids in laminar flow. *Trans. ASAE* 27:616.

CHAPTER 3

Convective Heat Transfer Coefficient Determination

3.1 INTRODUCTION

Heat transfer is defined as the transmission of energy from one region to another by means of a temperature gradient that exists between the two regions. Information on heat transfer is important in the thermal processing of foods. For instance, knowing the surface convective heat transfer coefficient is necessary in the study of food-processing applications such as blanching, retorting, freezing, and drying. This lab section is designed to give students hands-on experience in the evaluation of the heat transfer coefficient in some unit operations.

The principle involved in the calculations discussed in this chapter is the assumption of a negligible internal (conductive) heat transfer resistance of an object exposed to a convective heat transfer condition. For this assumption to be valid, the Biot number for an object immersed in a fluid should meet the following condition:

$$Bi = \frac{hL}{k} < 0.1 \qquad (3.1)$$

where

h = the convective heat transfer coefficient
k = the thermal conductivity of the solid material
L = the characteristic length of the object: (1) the radius for a sphere or a cylinder or (2) the thickness when heat is transferred from one side of

the plate or half the thickness when heat is transferred from both sides for a slab.

Because a small Biot number (Bi < 0.1) implies that the internal resistance to heat transfer is negligible, the energy (temperature) change measured at any point within the object is due to the net heat flow across the external heat transfer resistance. This statement is not valid when the Biot number is larger (Bi > 0.1) because the thermal resistance inside the object has to be taken into consideration. Therefore, a small aluminum object may be used to determine the external heat transfer coefficient because its internal resistance can be considered negligible due to high thermal conductivity.

There are several empirical relationships to compute the convective heat transfer coefficient when the Reynolds and Prandtl numbers are known. To obtain more information the student is advised to read References [1] and [2]. In the final part of this lab, the student will be asked to compare the experimental results with the predicted values obtained using these relationships and verify the validation of the assumption that the Biot number is lower than 0.1.

3.2 OBJECTIVES

(1) To determine the convective heat transfer coefficients of a small aluminum cylinder and a slab exposed to the following: (1) hot water, (2) boiling water, (3) steam blanching, (4) blast freezing, (5) still air freezing, and (6) tray dryer heating

(2) To compare the experimental data with predicted values obtained using available empirical equations

(3) To verify the assumption of a low Biot number and the linear relationship between temperature and time

3.3 MATERIALS AND METHODS

3.3.1 Materials

Aluminum models of a cylinder and a slab are required for the experiments described in this chapter. Aluminum was selected because it has a high thermal conductivity. The models have the same shapes or dimensions as real foods to ensure a useful estimation of the heat transfer coefficient in unit operations such as blanching, retorting, freezing, and drying.

The following equipment will be needed to obtain the surface heat transfer coefficient in several processing operations:

- a blast freezer (i.e., an Armfield blast freezer)
- a cold room at –40°C for a still air freezing situation
- a tray dryer (i.e., an Armfield tray dryer)
- a water bath for heat transfer in boiling water
- a steam retort
- an electronic balance to weigh the aluminum models
- a vernier caliper
- a PC with an appropriate interface card to record temperature history
- aluminum models with embedded thermocouples

3.3.2 Procedures

(1) Select a geometrically simple object (i.e., sphere, cylinder, or slab of aluminum) as the experimental model.
(2) Weigh and measure the dimensions of the model.
(3) Record the initial temperature of the model.
(4) Connect the thermocouples to the computer board.
(5) Select one unit operation to start the experiment and record the temperature every 2 sec. Keep doing this until minute temperature changes are observed with respect to time. It is expected that a small difference will always exist between the center of the object and the medium. Repeat this step for the other five processing situations mentioned above.
(6) Compute the convective heat transfer coefficient (see calculations below) for each situation.

3.3.3 Calculations

From an energy balance on the object, the change in internal energy is equal to the net heat flow from or to the object (cooling or heating). This can be written as (for a cooling process):

$$c_p V dT = hA(T - T_M)dtx \qquad (3.2)$$

where

c_p = specific heat of the object
V = volume of the object
ρ = density of the object

T = temperature of the object
A = surface area of the object
T_M = temperature of the medium
t = time

By integrating this expression between the limits of initial temperature T_0 to a temperature T, and from time 0 to time t, the following expression is obtained:

$$\ln \frac{T - T_M}{T_0 - T_M} = -\left(\frac{hA}{c_p \rho V} \right) t \qquad (3.3)$$

The left side of this equation is usually called the temperature ratio (TR) and a plot of log (TR) versus time results in a straight line with the following slope:

$$-\left(\frac{hA}{2.3 c_p \rho V} \right) \qquad (3.4)$$

The convective heat transfer coefficient can thus be determined if the Biot number is very small. The information needed to conduct these calculations is as follows:

- thermal conductivity of aluminum: 203 W/m K
- specific heat of aluminum: 0.896 kJ/kg K
- density of aluminum: 2707 kg/m³

3.4 EXPECTED RESULTS

(1) Compute the heat transfer coefficient for all the situations tested in the experiments. The use of a spreadsheet program (e.g., Excel®, Lotus 1-2-3®, Quatro®) is encouraged to present all original data and results. The format can be like that seen in Figure 3.1.
(2) Plot the corresponding semi-log graph of log (TR) versus time.
(3) Compute the slope of the line by using the least squares method of linear regression.
(4) Using the experimental heat transfer coefficient obtained in the lab, re-compute the Biot number for each situation. How close was the assumption of a Biot number less than 0.1? Did it hold for all cases? Why or why not?

Date:_____
Experiment: heat transfer coefficient determination on a blast freezer (brand name)

Time (t)	Temperature (T)	$TR\left(\ln\dfrac{T-T_M}{T_0-T_M}\right)$

Material tested: _____

Experimenter(s): _____

Initial temperature of model: _____

Weight of the model: _____

Dimensions of the model: _____

Notes:

FIGURE 3.1. Format example.

3.5 QUESTIONS

(1) Making reasonable assumptions with cited references, predict the heat transfer coefficient for each situation in the lab and compare the results with the experimental values obtained.

(2) What effect does an invalid assumption of a small Biot number ($Bi < 0.1$) have on heat transfer coefficient determination?

(3) Explain the difference in heat transfer coefficients obtained for the processes studied in the lab (e.g., hot vs. boiling water, blast vs. still air freezing) and discuss the implications that these processes will have on food quality (flavor, texture, nutrient retention, etc.).

3.6 REFERENCES

1. Heldman, D.R., and R.P. Singh. 1981. *Food Process Engineering.* 2nd ed. AVI. Westport, CT.
2. Perry, R.P., and D. Green. 1984. *Perry's Handbook of Chemical Engineering.* 6th ed. New York.
3. Kreith, F., and W.Z. Black. 1980. *Basic Heat Transfer.* Harper & Row Publishers, Philadelphia, PA.

CHAPTER 4

Thermal Processing of Foods:
Part I. Heat Penetration

4.1 INTRODUCTION

The study of heat penetration into foods is of great importance for a food engineer or scientist because heat processing is the most common technique used for food preservation today. Strict regulations and procedures are established by government agencies for the thermal processing of low-acid canned foods because there is widespread public health concern about the anaerobic *Clostridium botulinum,* a spore-forming microorganism that produces a toxin deadly to humans, even in very small amounts.

Two different methods for conventional thermal processing are known. In aseptic processing, the food product is sterilized prior to packaging. In canning, the product is packed and then sterilized. Although this laboratory experiment will focus on the canning process, the concepts applied here are also applicable to aseptic processing.

By doing the experiments described in this chapter, the student will become familiar with a retort operation and the basic parameters involved (e.g., heating and cooling rate, thermal lag). The next chapter will cover the experimental determination of the lethal effect of a thermal process on microorganisms of concern in food products.

To evaluate the effectiveness of a thermal process, it is necessary to know the thermal history of the product and the thermal resistance of the microorganism of concern. The first item will be covered in this chapter, and the second will be discussed in Chapter 5.

During a typical retort operation, the following three defined stages can be distinguished: (1) venting and come-up time, (2) cooking or processing cycle, and (3) cooling cycle. During venting, air inside the vessel is expelled, and the retort, depending on the design of the equipment, may reach the required processing temperature. If this is not the case, additional time should be allowed to obtain the processing temperature inside the retort. The processing cycle starts when the retort achieves the desired temperature, and the cooling cycle is at the end of the process, where pressurized water is typically used to avoid sudden expansion of the cans which may result in their damage.

4.2 OBJECTIVES

(1) To characterize the heating and cooling processes that a can undergoes during retorting when filled with different food products [The parameters to be measured in this process are the heating and cooling rate (f) and thermal lag (j).]
(2) To identify the cold point retort
(3) To verify the cold point inside the can, which will depend on the type of product being processed
(4) To determine the influence of the product on the heat penetration properties of the process

4.3 MATERIALS AND METHODS

4.3.1 Apparatus and Materials

The following materials and equipment, or their equivalent, are required to complete this laboratory exercise:

- canned raw material (It is necessary to choose some fluid-type food product, heated by convection inside a can, and a solid-type food heated by conduction. At least three different products are recommended for this lab to compare their heating characteristics.)
- a laboratory retort (with an automatic control for the venting time, processing temperature and time, and cooling cycle)
- thermocouples (installed inside a can with the appropriate connectors)
- a data logging system (a PC or equivalent is recommended with at least six channels)
- #303 or similar sized cans (suggested because they are commonly used for different types of food products)

- a can seamer
- a hole puncher tool, wrench, and a special screwing device for the type of thermocouples to be used

4.3.2 Procedures

This section is not intended to replace the retort operating instructions, but it contains the sequential steps for a successful and safe experiment. The student is expected to be familiar with the operation by reading the retort's operating manual before beginning the experiments.

4.3.2.1 FINDING THE COLD POINT OF THE RETORT

(1) Fill the retort completely with cans (if possible, full of product), and install six thermocouples in different places inside the retort.

(2) Close the door or lid and check that all closures on the retort are fastened securely.

(3) Check the data acquisition (temperature logger) system to ensure that it is working properly. Record the temperature of the six thermocouples at 1-min intervals.

(4) Open the vents and bleeders and close the drain and overflow.

(5) Admit steam by gradually opening both the controller and bypass lines.

(6) When the correct venting temperature has been reached and the specified time elapsed, close the vents. Using a table of thermodynamic properties of saturated steam for reference, examine the agreement between the retort's permanently installed mercury thermometer and pressure gauge readings as a criterion for the completion of air elimination. If the pressure gauge is high but the temperature is low, there is still some air in the retort, and venting should be continued until agreement is reached.

(7) Gradually close the bypass just before the processing temperature is reached to prevent a sudden drop in temperature which often occurs when the bypass is closed too rapidly.

(8) Maintain the retort temperature about one degree above the recommended processing temperature to help compensate for unavoidable fluctuations.

(9) As the process continues, check the temperature occasionally to make certain it is holding properly.

(10) Leave all bleeders wide open during the entire process.

(11) When the recommended time for the process has elapsed, turn off the steam and immediately start cooldown.

(12) Compare the temperature-time curve for six thermocouples, the lowest curve corresponding to the cold point of the retort. The temperature at this point will be used as the retort temperature (or *RT*, temperature of the steam) in the following experiment.

4.3.2.2 DETERMINING THE HEATING CURVE OF THE PRODUCT BEING PROCESSED

(1) Locate the materials for the heating curve determination experiments:
  - mushroom
  - tomato sauce
  - lunch meat
  - #303 cans
  - can seamer
  - thermocouples
  - data logging system
(2) Fill cans with the selected product (e.g., mushroom, tomato sauce, or lunch meat).
(3) Make sure the seaming machine is set for the can size you are working with.
(4) Turn the seamer and vacuum pump on, and then place both the can and lid in the machine, close the door, and pull the vacuum off. When a negative pressure of 15 in. of Hg has been reached, seal the can, release the vacuum, and take the can out.
(5) Determine the cold point in the product being processed keeping in mind that for a solid product, the coldest point will be at the center of the can, whereas in liquid-type products it will be on the lower part of the can axis (see Figure 4.1). Three or more thermocouple probes may be employed to

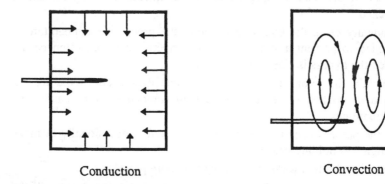

Conduction   Convection

FIGURE 4.1. Heat penetration to food in cans.

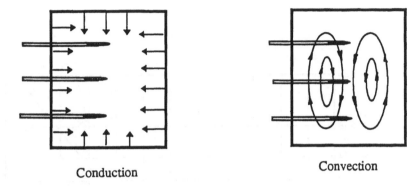

Conduction   Convection

FIGURE 4.2. Diagrams for the installation of three thermocouple probes.

locate the cold point (see Figure 4.2). If you are dealing with semisolid food, place several probes along the can axis to determine the critical point. Because all product temperature measurements will be done at this point, they will call the product temperature (PT).

(6) Collect the temperature data, and place the instrumented cans in the retort at the point you have chosen as the coldest. Connect the thermocouples to your data logger system, leaving one thermocouple to measure the temperature of the retort medium.

(7) Select the venting and processing time and processing temperature for the retort. For the venting time, a minimum of 3 min is advisable to eliminate all the air inside the vessel. The processing temperature should be somewhere between 240 and 260°F, which is normal for this type of process. The processing (or cooking) time indicated in the control cabinet is the time elapsed between opening and shutting off the steam. A good starting point is somewhere between 10 and 15 min.

(8) Start the retort again and record your *PT* and *RT* at 5-sec intervals. Although you have already set your processing time, it can be extended or shortened until a temperature difference of about 2 to 5°F between PT and RT is reached.

4.3.3 Calculations

The typical heat penetration curve is shown in Figure 4.3. The governing equation for either a cooling or heating process can be written [2] as follows:

$$\log\left(RT - PT\right) = -\left(\frac{t}{f_h}\right) + \log\left(RT - T_{pih}\right) \qquad (4.1)$$

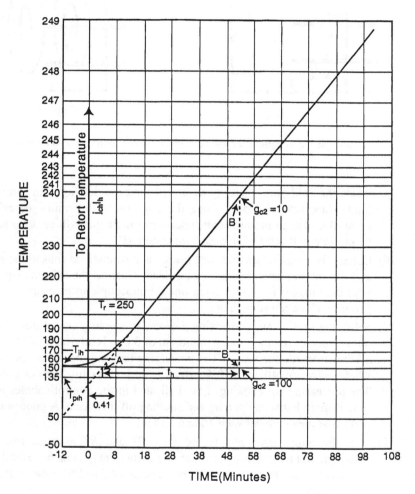

FIGURE 4.3. Heat penetration curve (reprinted from Reference [3], p. 139, © 1973, Academic Press).

where

PT = product temperature
RT = retort temperature
T_{pih} = pseudo initial heating temperature
t = time
f_h = heating rate (f_c: cooling rate)

The value j is called the thermal lag (heating or cooling lag) and can be defined for either the heating or cooling process as [1]:

$$j_h = \frac{RT - T_{pih}}{RT - T_{ih}}$$ (4.2)

where T_{ih} is the temperature at the beginning of the heating process. Clearly the value f_h (h means heating, and the equivalent f_c can be obtained for cooling) is the slope of the linear part of a plot log temperature versus time and is defined as the time in minutes required for the heat penetration curve to traverse one log cycle.

If the difference between RT and initial PT is called I_h, Equation (4.1) can be written as:

$$\log (RT - PT) = -\left(\frac{t}{f_h}\right) + \log (j_h I_h)$$ (4.3)

Notice that theoretically the PT will never reach the RT. Hence, this finite difference in temperature at the end of the heating process is defined as g, and the process time as B. Equations (4.1) and (4.3) can therefore be written as:

$$\log (g) = -\left(\frac{B}{f_h}\right) + \log (j_h I_h)$$ (4.4)

or

$$B = -f_h \log \left(\frac{j_h I_h}{g}\right)$$ (4.5)

Some food products, such as those containing a significant amount of starch, show a broken-type heating curve as shown in Figure 4.4. This situation will be studied in detail in Chapter 5.

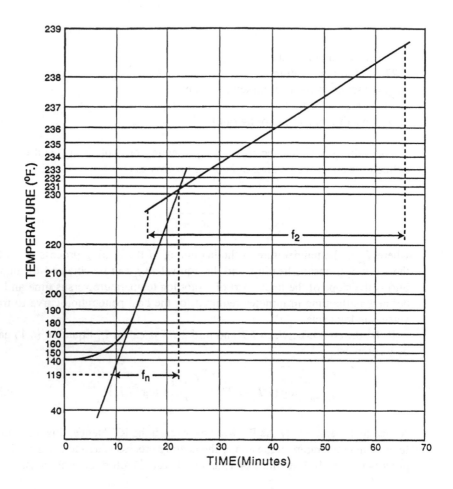

FIGURE 4.4. Broken line-type heating curve (reprinted from Reference [3], p. 162, © 1973, Academic Press).

4.4 EXPECTED RESULTS

(1) Draw a plot of temperature versus time for several points in the retort and find the location of the cold point in the retort.

(2) Draw a plot of log *PT* minus *RT* against time. Observe whether you have a single or broken line relationship in your plot.

(3) Obtain the characteristic parameters *f* and *j* for the heating and cooling stages of the process. In the case of a broken line-type curve, refer to

Figure 4.2 to compute the parameter $f_h 2$. Use a linear regression analysis of your data to obtain the required parameters.

4.5 QUESTIONS

(1) How is the f value affected by the physical properties of the food?
(2) How is the j value affected by the physical properties of the food?
(3) Draw a plot of temperature versus time of the entire retort operation which clearly identifies the processing, heating, cooling, and total times.
(4) How is the cold point in the product affected by the physical characteristics of the food? Which physical property do you think most affects the heating (or cooling) behavior?
(5) What would you expect to be the coldest point in the retort? Why?
(6) Do the experimental results agree with your expectations? If not, why?
(7) Why is venting important before and during the retort operation?
(8) List what kind of food products may exhibit a broken line-type heating curve. Why?

4.6 REFERENCES

1. Karel, M., Fennema, O.R., and D.B. Lund. 1975. *Physical Principles of Food Preservation.* Marcel Dekker, New York.
2. Heldman, D.R. and R.P. Singh. 1981. *Food Process Engineering.* AVI, Westport, CT.
3. Stumbo, C.R. 1973. *Thermobacteriology in Food Processing,* 2nd ed. Food Science and Technology series. Academic Press, New York.

CHAPTER 5

Thermal Processing of Foods: Part II. Lethality Determination

5.1 INTRODUCTION

In Chapter 4, the problems of characterizing the process of heat penetration into foods were studied, with the concepts of heating rate and thermal lag defined and measured in the lab. We will now focus on the problem of relating the lethal effect on the microorganism of interest when processing a food product.

The term *lethality* refers to microbial inactivation. To understand the kinetics of microbial or enzymatic inactivation, some background is provided below. The reader interested in obtaining more detailed information may refer to the classical book by Stumbo [1].

When the heat transfer mechanism is dominated by convection in a thermal process that is liquid or highly viscous, there is movement of the product inside the container due to natural convection and buoyancy forces. It is assumed that every point in the container receives an integrated lethal effect at least as large as that received by the cold point. (Remember from Chapter 4 that the cold point for a convection heating product is located near the bottom of the container's axis.)

When the heat transfer mechanism is dominated by conduction, there is no product movement inside the can. During either the heating or cooling process there will be a temperature gradient from the wall to the center of the container, which is considered the cold point even though it is not necessarily the least lethal point throughout the cross section of the can. To better understand this, imagine that at the end of the heating process the center of the can is at a temperature below the retort temperature (RT) when cooling starts. At this time, even though the wall is being rapidly cooled, the center still "sees" a thermal driving force, which means the cold point will be held at a higher temperature considerably longer than other parts of the can, making a significant difference in the lethal effect that all single points in the container receive.

The solution to the problem of finding an integrated lethality for the whole can is quite complex. Several approaches have been developed, some requiring the use of computers. It has been found that the location of the least lethal point in the container is a function of container geometry and process conditions.

The purpose of this laboratory exercise is to understand and evaluate a thermal process in a food product heated by conduction and convection. In the former, the approach by Stumbo [2] will be followed. Inside the can several imaginary surfaces, typically called iso-j or iso-F, can be conceived as receiving the same thermal treatment. The shape of these surfaces in a cylindrical can is also cylindrical near the wall, and they change their shape to ellipsoids and spheres as they approach the center of the container. Stumbo also found that the difference between the lethal value received by any point on these surfaces (F_s) and the lethal value received by the center of the can (F_c) shows a linear relationship with a fraction of the total volume enclosed by the iso-j. Stumbo developed his procedure based on this fact, doing a linear interpolation to obtain a F_c value from a desired integrated lethality value (F_s). Once the F_c value is known, the procedure is the same as in the case of convection heating products. (This method is explained in detail in Section 5.3.3.) To avoid cumbersome computations in determining the integrated lethality of a thermal process in a solid food product, some tables have been published [2] listing the F_c value as a function of the heating characteristics of the process for every commercial can size.

5.2 OBJECTIVES

(1) To understand the physical principles affecting thermal lethality of microorganisms in foods

(2) To evaluate the thermal lethality in a food system given a specific thermal process by using both the improved general (graphic) and formula methods (analytical)

(3) To become familiar with a typical food-processing problem consisting of the determination of a suitable processing time for a new canned food product

(4) To compute the integrated lethal effect in the food due to a specific heat treatment

(5) To measure the influence of the size of the container as well as the processing conditions in the lethality of a thermal process

5.3 MATERIALS AND METHODS

5.3.1 Apparatus and Materials

Two kinds of raw materials are required for both convection and conduction heating:

- lunch meat (ground pork or tuna)
- mushrooms, green beans, olives

Students are encouraged to make suggestions about different products they want to test. A low-acid food is preferred because it has great commercial importance in a sterilization process due to the possible presence of *Clostridium botulinum*. Other materials required for this lab are described in Chapter 4.

5.3.2 Procedures

(1) Locate two kinds of raw materials (e.g., mushrooms and ground beef or other appropriate food).
(2) Prepare three cans for each raw material with a thermocouple installed at the bottom, and then fill them with raw material and seal using the can seamer. (For details, see Chapter 4.)
(3) Obtain the heat penetration data for the cans following the guidelines in Chapter 4.
(4) Process the cans until the temperature of the cold point is within 1° or 2°F of the retort temperature (*RT*).
(5) Process all cans following the procedure described in the previous two chapters. Each can size and product must be processed selecting three different retort temperatures. The student may choose 245, 260, or 275°F and collect heat penetration data for each can (two products, three processing temperatures, and three replicates).

5.3.3 Calculations

Before starting with the typical calculations for lethality determination, the following terms need to be understood:

5.3.3.1 SOME DEFINITIONS

D Decimal Reduction Time is the time required to reduce the population of microorganisms by 90% when processed at a specific temperature.
TDT The plot of log thermal death time (minimum time at a specific temperature to accomplish total destruction), represented as TDT_T against

temperature. Because the rate of microbial inactivation follows a first-order kinetics, this plot results in a straight line. Figure 5.1 shows a typical *TDT* curve with a slope of $-1/z$.

The time required at a specific temperature to achieve a certain degree of microbial inactivation for a specific product and a value of lethality obtained after a certain thermal processing. It is specific for each microorganism at a fixed temperature. For *Clostridium botulinum*, the F value represents the heating time required to reduce the population by a factor of 10^{12}. A common way to represent an *F* value is by writing the reference temperature as a subscript and the z value (see below) of the microorganism as a superscript, [e.g. F_{250}^{18} represents the *F* value for *Clostridium botulinum* ($z = 18°F$) at 250°F and

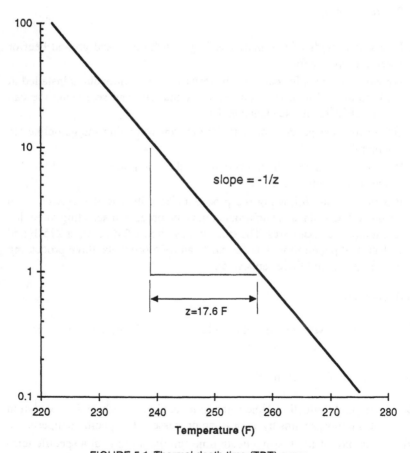

FIGURE 5.1. Thermal death time (TDT) curve.

F_{250}^{14} represents the F value for *Bacillus stearothermophilus* ($z = 14°F$) at 250°F]. The superscripts and subscript are omitted most of the time for simplicity and the lethal effect equivalent to 1 min is represented as F_t.

z The increase in temperature (°F) required to reduce the thermal death time tenfold in a *TDT* plot (see Figure 5.1). The z value measures the effect of temperature on the reaction rate constants. The following equation describes a *TDT* curve:

$$\log\left(\frac{F_T^z}{F_{250}^z}\right) = \frac{250 - T}{z} \tag{5.1}$$

T_{ih} The initial temperature of the food at the beginning of the heating (when steam enters the retort) or cooling process (see Chapter 4).

T_{pih} The pseudo initial heating temperature resulting from the intersection of the extension of the straight portion of the semi-log heating curve and a vertical line at the beginning of the process.

T_r The temperature of the heating medium (e.g., steam) in the retort.

B The time it takes for the retort to reach the desired temperature until the steam is turned off, plus 40% of the time required to bring the retort to the required processing temperature. B must be distinguished from the so-called operator time (P_t), which is the time accounted for from the instant the retort reaches the processing temperature until the steam is turned off.

$$B = P_t + 0.4\ell \tag{5.2}$$

I_h The difference between retort and initial food temperature at the beginning of the heating process.

$$I_h = T_r - T_{ih} \tag{5.3}$$

I_c Analogous to I_h, this is the difference between the retort and food temperature at the beginning of the cooling process.

g The difference between the retort and maximum product temperature measured at the cold point.

U The time required at RT to accomplish the same amount of bacterial destruction as in the F value of the process.

The heating and cooling factors and the heating and cooling rates (f_h and f_c) are defined in Chapter 4. The relationship between U and F_t is:

$$U = F \cdot F_i = F \cdot 10^{\frac{250-T}{z}} \tag{5.4}$$

5.3.3.1.1 Graphical Method

From the equation of a *TDT* curve [Equation (5.1)] the ratio is:

$$\frac{F_T^z}{F_{250}^z} = 10^{\frac{250-T}{z}} = \frac{TDT}{1} \tag{5.5}$$

Notice the unity value assigned to 250°F as the denominator of the right term in this equation. This means that the lethality effect of heating 1 min at 250°F has been assigned, and the *F* value is equal to unity (unit sterility).

It can be inferred from Equation (5.5) that the ratio of the *F* value at 250°F over the *F* value at any processing temperature is no more than the inverse of the *TDT* at that temperature. This 1/*TDT* is the sterility rate and can be computed easily using Equation (5.5). For example, supposing a food product is at 235°F and has a *z* value of 14. The sterility rate is compared as:

$$\frac{1}{TDT_{235}} = 10^{\frac{235-250}{14}} = 0.0849$$

This means 1 min at 235°F is equal to 0.0849 min at 250°F. Another way to look at the same situation is:

$$\frac{TDT_{235}}{1} = 10^{\frac{250-235}{14}} = 11.78$$

which means that 11.78 min at 235°F is equivalent to 1 min at 250°F.

The integrated lethality may be obtained from a graphical integration procedure. The heat penetration data is converted to lethal rates (1/*TDT*) by using Equation (5.5). From a plot of lethal rate against time, the area under the curve can be obtained by appropriate computer software. The area is computed in "units sterility area" (USA), which represents a process equivalent to 1 min at 250°F (*F* = 1).

It is improbable that the experimental time selected will be the same as the process time required for the desired *F* value in an industrial operation. The right process time can be obtained by drawing several lines parallel to the cooling curve and interpolating the process time required to obtain a predetermined *F* value.

5.3.3.2 FORMULA METHOD

To establish the required F value for the process, the following model can be used:

$$F = D \cdot \left[\log (a) - \log (b) \right] \tag{5.6}$$

where

$D = D$ value of the product evaluated at reference temperature
$a =$ initial load of microorganisms
$b =$ desired final load of microorganisms (when less than 1, it is the probability of finding one microorganism per container or gram)

The U value is required for this calculation and can be obtained from Equation (5.4). To obtain the value B of processing time in Equation (4.5), the value g must be evaluated. Tabulated $f_h/U:g$ are available in Table 5.1, and more complete tables in Reference [1]. It has been found that this relation is dependent on the cooling driving force, which is a function of the cooling lag factor j_c.

Tables are normally given for $I_c + g = RT - CT$ (retort temperature minus cooling water temperature) equal to 180°F, which is the situation normally found. Corrections can be made to the actual F value obtained by subtracting 1% from the theoretical value for each 10°F, where the $I_c + g$ is above 180°F, and by adding 1% for each 10°F below 180°F.

In the case of products exhibiting one straight line after the initial lag, the process temperature B can be obtained from:

$$B = f_h \left[\log \left(j_h I_h \right) - \log (g) \right] \tag{5.7}$$

In the case of a broken line (see Figure 4.4), the following relationship can be applied:

$$F = \frac{f_2}{\dfrac{f_h}{U_2} F_i} - \frac{r \left(f_2 - f_h \right)}{\dfrac{f_h}{U_h} F} \tag{5.8}$$

5.3.4 Example 1

To illustrate a typical calculation, demonstrate a thermal process by inactivating *Clostridium sporogenes* using the following data:

- F value required for the process: 9.8 min
- *RT*: 240°F

TABLE 5.1. f_h/U:g Relationships when z = 18.[a]

f_h/U	Values of g when j of Cooling Curve is:								
	0.40	0.60	0.80	1.00	1.20	1.40	1.60	1.80	2.00
0.20	4.09–05[b]	4.42–05	4.76–05	5.09–05	5.43–05	5.76–05	6.10–05	6.44–05	6.77–05
0.30	2.01–03	2.14–03	2.27–03	2.40–03	2.53–03	2.66–03	2.79–03	2.93–03	3.06–03
0.40	1.33–02	1.43–02	1.52–02	1.62–02	1.71–02	1.80–02	1.90–02	1.99–02	2.09–02
0.50	4.11–02	4.42–02	4.74–02	5.06–02	5.38–02	5.70–02	6.02–02	6.34–02	6.65–02
0.60	8.70–02	9.43–02	1.02–01	1.09–01	1.16–01	1.23–01	1.31–01	1.38–01	1.45–01
0.70	0.150	0.163	0.176	0.189	0.202	0.215	0.228	0.241	0.255
0.80	0.226	0.246	0.267	0.287	0.308	0.328	0.349	0.369	0.390
0.90	0.313	0.342	0.371	0.400	0.429	0.458	0.487	0.516	0.545
1.00	0.408	0.447	0.485	0.523	0.561	0.600	0.638	0.676	0.715
2.00	1.53	1.66	1.80	1.93	2.07	2.21	2.34	2.48	2.61
3.00	2.63	2.84	3.05	3.26	3.47	3.68	3.89	4.10	4.31
4.00	3.61	3.87	4.14	4.41	4.68	4.94	5.21	5.48	5.75
5.00	4.44	4.76	5.08	5.40	5.71	6.03	6.35	6.67	6.99
6.00	5.15	5.52	5.88	6.25	6.61	6.98	7.34	7.71	8.07
7.00	5.77	6.18	6.59	7.00	7.41	7.82	8.23	8.64	9.05
8.00	6.29	6.75	7.20	7.66	8.11	8.56	9.02	9.47	9.93
9.00	6.76	7.26	7.75	8.25	8.74	9.24	9.74	10.23	10.73
10.00	7.17	7.71	8.24	8.78	9.32	9.86	10.39	10.93	11.47
15.00	8.73	9.44	10.16	10.88	11.59	12.31	13.02	13.74	14.45
20.00	9.83	10.69	11.55	12.40	13.26	14.11	14.97	15.82	16.68
25.00	10.7	11.7	12.7	13.6	14.6	15.6	16.5	17.5	18.4
30.00	11.5	12.5	13.6	14.6	15.7	16.8	17.8	18.9	19.9
35.00	12.1	13.3	14.4	15.5	16.7	17.8	18.9	20.0	21.2

TABLE 5.1. continued

f_H/U	0.40	0.60	0.80	1.00	1.20	1.40	1.60	1.80	2.00
40.00	12.8	13.9	15.1	16.3	17.5	18.7	19.9	21.1	22.3
45.00	13.3	14.6	15.8	17.0	18.3	19.5	20.8	22.0	23.2
50.00	13.8	15.1	16.4	17.7	19.0	20.3	21.6	22.8	24.1
60.00	14.8	16.1	17.5	18.9	20.2	21.6	22.9	24.3	25.7
70.00	15.6	17.0	18.4	19.9	21.3	22.7	24.1	25.6	27.0
80.00	16.3	17.8	19.3	20.8	22.2	23.7	25.2	26.7	28.1
90.00	17.0	18.5	20.1	21.6	23.1	24.6	26.1	27.6	29.1
100.00	17.6	19.2	20.8	22.3	23.9	25.4	27.0	28.5	30.1
150.00	20.1	21.8	23.5	25.2	26.8	28.5	30.2	31.9	33.6
200.00	21.7	23.5	25.3	27.1	28.9	30.7	32.5	34.3	36.2
250.00	22.9	24.8	26.7	28.6	30.5	32.4	34.3	36.2	38.1
300.00	23.8	25.8	27.8	29.8	31.8	33.7	35.7	27.7	39.7
350.00	24.5	26.6	28.6	30.7	32.8	34.9	37.0	39.0	41.1
400.00	25.1	27.2	29.4	31.5	33.7	35.9	38.0	40.3	42.3
450.00	25.6	27.8	30.0	32.3	34.5	36.7	38.9	41.2	43.4
500.00	26.0	28.3	30.6	32.9	35.2	37.5	39.8	42.1	44.4
600.00	26.8	29.2	31.6	34.0	36.4	48.8	41.2	43.6	46.0
700.00	27.5	30.0	32.5	35.0	37.5	39.9	42.4	44.9	47.4
800.00	28.1	30.7	33.3	35.8	38.4	40.9	43.5	46.0	48.6
900.00	28.7	31.3	34.0	36.6	39.2	41.8	44.4	47.0	49.7
999.99	29.3	31.9	34.6	37.3	39.9	42.6	45.3	47.9	50.6

aReprinted from Reference [1], p. 260, © 1973, Academic Press.
b4.09–05 means 4.09×10^{-5}.

- z value: 18°F
- heating rate f_h: 14 min
- heating lag factor j_h: 1.0
- cooling lag factor j_c: 1.0
- initial food temperature: 140°F

5.3.4.1 SOLUTION

To obtain U, the value of F_i must be computed:

$$F_i = 10^{\frac{250-T}{z}} = 10^{\frac{250-240}{18}} = 3.593 \text{ min}$$

$$U = F \cdot F_i = 9.8 \cdot 3.593 = 35.22 \text{ min}^2$$

the value f_h/U is computed as follows:

$$\frac{f_h}{U} = \frac{14}{35.22} = 0.4 \text{ min}$$

The corresponding g value is obtained from Table 5.1:

$$g = 1.62 \cdot 10^{-2}$$

To compute the process time B, find I_h:

$$I_h = T_r - T_{ih}$$

$$I_h = 240 - 140 = 100 \text{ min}$$

Now compute the process time B:

$$B = f_h \left(\log j_h I_h \log g \right)$$

$$B = 14 \left[\log \left(1 \cdot 100 \right) - \log \left(1.62 \cdot 10^{-2} \right) \right] = 53.06 \text{ min}$$

5.3.5 Example 2

Obtain the processing time by using the calculation for conduction heating, provided the required integrated lethality is known.

5.3.5.1 SOLUTION

To establish the desired F value for the process (F_s), the following expression can be used:

$$F_s = D_{250}(\log a - \log b) \tag{5.9}$$

where

$D_{250} = D$ value at 250°F
    a = initial microorganism load
    b = final microorganism load (When b is less than unity, it can be thought of as the probability of finding one microorganism per container.)

The following expression relates the F_s value with the F value at the center of the container (F_c):

$$F_s = F_c + D_{250}\left[1.084 + \log\left(\frac{F_\lambda - F_c}{D_{250}}\right)\right] \tag{5.10}$$

in which F_λ characterizes the heat treatment received by any iso-j region.

The problem now is to obtain an F_c desired from F_s. To do this, a linear interpolation process proposed by Stumbo can be used. Two initial values of F_c are selected, named F_{c1} and F_{c2}. From there, two F_s values are computed (F_{s1} and F_{s2}) using Equation (5.10). The corresponding $F_{\lambda 1}$ and $F_{\lambda 2}$ can be obtained from the following relationships:

$$g_\lambda = 0.5\, g_c \tag{5.11}$$

$$j_\lambda = 0.5\, j_c \tag{5.12}$$

The F_c value can be obtained from the known parameters F_{c1}, F_{c2}, F_{s1}, F_{s2} by doing the interpolation below:

$$F_c = F_{c1} + \frac{F_{c2} - F_{c1}}{F_{s2} - F_{s1}}\left(F_s - F_{s1}\right) \tag{5.13}$$

To choose the starting values, take into account that F_c is always less than F_s, and the difference between the F_s and F_c desired is generally more than one D_{250} value. The starting values should be closer to the desired F_c, but one should be higher and the other lower.

5.3.6 Example 3

To illustrate how to obtain good starting values for F_{c1} and F_{c2}, assume the procedure for the inactivation of *Bacillus stearothermophillus*.

- z: 14°F
- D_{250}: 0.2 min
- F_s required: 3.01 min

Hence, F_c is likely to be lower than 2.81 (3.01 − 0.2), making $F_{c1} = 2$ and $F_{c2} = 3$ likely starting values. After the interpolation is obtained, the F_c value should be between the starting values. Once the F_c value is obtained, the single point analysis (as in convection heating products) can be done. To simplify these cumbersome calculations the reader should refer to Table 5.2 in which the values for F_c and F_s can be read for certain f_h, z, and T_{th} values and a specific can size, but be sure that the information is available for the can size used in the experiments. Do not try to extrapolate any value from Table 5.2 because it usually leads to erroneous results.

5.4 EXPECTED RESULTS

(1) Obtain the integrated lethality effect (F value) for each of the products processed by using both the graphical and formula methods. For the former, any spreadsheet software (i.e., Excell®, Lotus 1-2-3®) may be used to perform the calculations and print the graphs. Be sure to submit all the calculations used.

(2) Obtain the equivalent lethality for all the cans processed in the laboratory. There should be two can sizes filled with two products and processed at three different temperatures. The report should include the heat penetration curves and standard parameters f_h, f_c, j_h, and j_c.

5.5 QUESTIONS

(1) Assume that you now need to process a fluid food product in which *Coxiella burnetti* ($z = 9.2$ and $D_{150} = 0.55$) is your microorganism of interest. The initial population is 55,000 microorganisms/gram and the desired final load 1 microorganism/gram. If the product is going to be canned in a #303 can, determine the processing temperature and time it takes to achieve the desired conditions. (Assume the heat penetration data are available and the same as that collected during the actual experiments in the lab.)

TABLE 5.2. Process Time (B) at Retort Temperature of 250°F for Convection Heated Foods in a 303 × 406 Can.[a]

f_h	z	100	110	120	130	140	150	160	170	180	190	200	Equivalent $F_c^{18} = F_s^{18}$
2.66	12	7.2	7.1	7.0	6.9	6.8	6.7	6.6	6.5	6.3	6.1	5.9	3.30
	14	7.0	7.0	6.9	6.8	6.7	6.6	6.4	6.3	6.2	6.0	5.8	3.13
	16	6.9	6.8	6.7	6.6	6.5	6.4	6.3	6.2	6.0	5.8	5.6	3.00
	18	6.8	6.8	6.7	6.6	6.5	6.4	6.2	6.1	6.0	5.8	5.6	2.94
5.33	12	11.3	11.1	10.9	10.8	10.5	10.3	10.1	9.8	9.5	9.1	8.7	3.51
	14	11.0	10.8	10.6	10.5	10.3	10.0	9.8	9.5	9.2	8.9	8.4	3.28
	16	10.7	10.5	10.3	10.2	10.0	9.7	9.5	9.2	8.9	8.6	8.1	3.06
	18	10.5	10.3	10.2	10.0	9.8	9.6	9.3	9.0	8.7	8.4	7.9	2.94
7.99	12	15.0	14.8	14.5	14.2	13.9	13.6	13.3	12.8	12.4	11.8	11.2	3.93
	14	14.6	14.3	14.1	13.8	13.5	13.2	12.8	12.4	11.9	11.4	10.8	3.55
	16	14.1	13.9	13.6	13.3	13.0	12.7	12.3	11.9	11.4	10.9	10.3	3.19
	18	13.7	13.5	13.2	12.9	12.6	12.3	11.9	11.5	11.1	10.5	9.9	2.94
10.65	12	18.7	18.4	18.0	17.7	17.3	16.8	16.3	15.8	15.2	14.5	13.6	4.19
	14	18.0	17.7	17.4	17.0	16.6	16.2	15.7	15.1	14.5	13.8	13.0	3.76
	16	17.4	17.1	16.7	16.4	16.0	15.5	15.0	14.5	13.9	13.2	12.3	3.32
	18	16.8	16.5	16.1	15.8	15.4	14.9	14.4	13.9	13.3	12.5	11.7	2.94
13.31	12	22.3	21.9	21.4	21.0	20.5	19.9	19.3	18.6	17.8	17.0	15.9	4.51
	14	21.4	21.0	20.6	20.1	19.6	19.1	18.4	17.8	17.0	16.1	15.1	3.93
	16	10.6	20.2	19.7	19.3	18.8	18.2	17.6	16.9	16.2	15.3	14.2	3.42
	18	19.7	19.3	18.9	18.4	17.9	17.4	16.8	16.1	15.3	14.4	13.4	2.94

Initial Food Temperature (T_{ih})

[a]Reprinted from Reference [2], p. 96. © 1983, Academic Press.

(2) Explain the effect the size of the container will have on the lethality of the thermal process.

(3) Explain how the viscosity (in case of a Newtonian fluid) will affect the lethality of a given thermal process.

(4) Compute the required processing time to accomplish a 12 log cycle reduction in one of the products used in the lab, assuming *Clostridium botulinum* ($z = 18°F$) is the microorganism of greatest concern. The initial bacterial population is 720 microorganism/cm^3. Use one can size and processing temperature tested in the laboratory.

(5) Now assume *Bacillus stearothermophillus* ($z = 14°F$) is the microorganism of concern, and compute the required processing time assuming the same conditions as noted in Question 4.

(6) With one set of data available from the experiments, assume that your product is convection heated. Compute the required processing time to achieve the same lethality you obtained for this situation. Compare the results obtained using both analyses. Did you find any difference? Explain.

(7) In which case (convection or conduction heating) is nutrient retention favored? Why?

5.6 REFERENCES

1. Stumbo, C.R. 1973. *Thermobacteriology in Food Processing.* 2nd ed. Academic Press, New York.
2. Stumbo, C.R., K.S. Purohit, T.V. Ramakrishan, D.A. Evans, and F.J. Francis. 1983. *Handbook of Lethality Guides for Low-Acid Canned Foods.* CRC Press, Boca Raton, FL.
3. Karel, M., O.R. Fennema, and D.B. Lund. 1975. *Physical Principles of Food Preservation.* Marcel Dekker, New York.

CHAPTER 6

Freezing of Foods

6.1 INTRODUCTION

Freezing, one of the most common processes for the preservation of food, is effective for retaining the flavor, color, and nutritive value of food and is moderately effective for the preservation of texture. However, solid foods from living tissues such as meat, fish, fruits, and vegetables are of cellular structure with delicate cell and cell membranes. Water, within and between cells, forms minute ice crystals when rapidly frozen. When it freezes slowly, large ice crystals and clusters of crystals develop, causing much more physical rupture and separation of cells than small crystals. Large ice crystal damage is detrimental not only to cellular foods but can also disrupt emulsions such as butter, frozen foams such as ice cream, and gels such as pudding and pie fillings.

Compared with the thermal processing of foods, preservation by freezing is not based on microbial destruction but on the following two main mechanisms: (1) the decrease in enzyme and microbial activity that is a result of the low temperatures involved in the freezing process and (2) the conversion of water into ice, which reduces the availability of water activity for enzymatic and other degradative reactions in the food system.

In the study of freezing, many theories have been developed, and the assumptions used for each approach greatly differ from one another. For more information on these theories, refer to References [1] and [2]. From a food engineer's point of view, three topics are of considerable interest in the study of freezing:

(1) Determination of the initial freezing point of a food product. This will be different from the freezing point of pure water because of the presence of solutes in the food that reduce the freezing point: The basic colligative

property of solutions causes increases in solute concentrations to lower the freezing point from that of pure solvent. Thus, the more salt, sugar, minerals, or proteins in a solution, the lower its freezing point and the longer it will take to freeze when put into a freezing chamber. Because different foods can have quite different water contents and type and amount of dissolved solids, they have different initial freezing points, and under a given freezing condition, require different times to reach a solidified frozen state.

(2) Determination of the freezing time required for a food system: Familiarity with the pattern of temperature changes food materials undergo during freezing is basic to an understanding of the process. A given unit of food will not freeze uniformly (i.e., it will not suddenly change from liquid to solid). In the case of milk placed in the freezer, the liquid nearest the container walls will freeze first, and these initial ice crystals will be pure water. As water continues to be frozen out, the milk will become more concentrated in minerals, proteins, lactose, and fat. This concentrate, which gradually freezes, also becomes more concentrated as freezing proceeds. Finally, a central core of highly concentrated unfrozen liquid remains, and if the temperature is sufficiently low, it too will ultimately freeze.

(3) Determination of the energy requirements to achieve the degree of freezing necessary for a food product to properly select the refrigeration equipment to be used: This lab will address only the first two points. To study the first, several relationships have been developed. We will follow the Schwartzberg approach [1] in which the freezing point depression is expressed as a function of the fractions of solutes in a solution (n_s) and the mass of water bound per unit mass of solute (b). For the freezing time determination, the equation by Planck, with further modification by Nagaoka, will be presented.

Experimental verification of the heat requirements for freezing requires the use of a differential scanning calorimeter (DSC), which is out of the scope of this laboratory practice.

6.2 THEORY

The presence of solute in a food reduces the freezing point of the water contained in the system. The following relationship by Schwartzberg [1] is based on the assumption that water activity is equal to the mole fraction of the solvents in a food system (Raoult's law), which gives an approximation with 2% of experimental data for ideal and dilute solutions.

$$\frac{En}{n_w - bn_s} \approx \frac{18.02\Delta H_0 \left(T_0 - T\right)}{RT_0^2}$$

$$= 0.009678\left(T_0 - T\right) \qquad (6.1)$$

where

$E = 18.02/M_s$

M_s = molecular weight of the solute

n_w = mass fraction of water

n_s = mass fraction of solute

b = mass of water bound per unit mass of solutes and solids (in the case of sucrose solution, $b = 0.21$; for glucose $b = 0.17$, and for coffee extract, $b = 0.371$)

T = freezing temperature in K

T_0 = 273.16 K

R = universal gas constant (8.314×10^3 kg m^2 sec^{-2} kg-mol^{-1} K^{-1})

ΔH_0 = latent heat of fusion of ice at T_0

To estimate the freezing time, the following expression by Planck may be used:

$$t_F = \frac{\rho\Delta H}{T_F - T_\infty}\left(\frac{Pa}{h_c} + \frac{Ra^2}{k}\right) \qquad (6.2)$$

where

t_F = freezing time

T_F = initial freezing point

T_∞ = medium temperature

h_c = convective heat transfer coefficient

k = thermal conductivity of food

a = thickness of a slab if heat is transferred to one surface; half thickness if both surfaces (radius of a cylinder, the smallest dimension of the geometry)

ρ = density of the food (unfrozen)

P, R = empirical constant for:

- an infinite long slab: $P = 1/2$, $R = 1/8$
- an infinite long cylinder: $P = 1/4$, $R = 1/16$
- a sphere: $P = 1/16$, $R = 1/24$
- a brick or block geometry: P and R are obtained from Figure 6.1

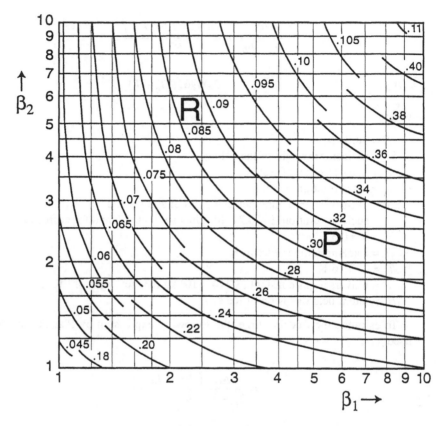

FIGURE 6.1. P and R to be used in the Planck equation for a geometry (reprinted from Reference [2], p. 180, with permission of Chapman & Hall).

To use the chart in Figure 6.1, the values β_1 and β_2 need to be computed. Heldman and Singh [2] defined β_1 as the multiplying factor required to make the product of β_1 and the smallest dimension (a) equal to the second smallest dimension of the geometry. Similarly, β_2 is the factor that makes the product of β_2 and the smallest dimension (a) equal to the largest dimension of the geometry.

Despite its usefulness, Planck's equation has major limitations. For instance, it assumes that all the water freezes at one temperature, the latent heat of fusion is constant during the freezing process, and that there is a constant thermal conductivity for the frozen region. Notice also that the initial and final temperature of the product are not taken into account in the equation. To overcome some of these drawbacks, Nagoaka proposed a modification to Planck's original equation:

$$t_F = \frac{\rho \Delta H'}{T_F - T_\infty}\left(\frac{Pa}{h_c} + \frac{Ra^2}{k}\right)$$

where

$$\Delta H' = \left[1 + 0.00445(T_0 - T_F)\right]\left[c_{pU}(T_i - T_F) + \Delta H + c_{pI}(T_F - T)\right]$$

and

T_F = desired final temperature of the product
T_0 = initial temperature
c_{pU} = specific heat of a food at its initial unfrozen temperature
c_{PI} = specific heat of ice

6.3 OBJECTIVES

(1) To obtain the freezing curves of several food products (e.g., fruit juice, apple slices, beef) and compare them with the freezing curve of pure water
(2) To verify the theoretical freezing point for two sugar solutions in water
(3) To obtain the freezing time required under still and blast freezing using the approaches described by Planck and Nagaoka

6.4 MATERIALS AND METHODS

6.4.1 Materials

The following materials or their equivalent are necessary for this lab:

- two freezing mediums used for blast and still air freezing (An example of the first is the Armfield model FT 35. For still air freezing, a cold room at −40°C or an equivalent big freezer can be used, but protection from evaporator fan produced air movement is necessary.)
- sucrose, glucose, and sodium chloride to prepare solutions that will illustrate the freezing point depression phenomenon
- distilled water to be used as a reference material
- 250-ml volumetric flasks
- metallic or plastic containers (e.g., 250-ml plastic beakers) to deposit and freeze liquid materials
- food products to be frozen (i.e., orange juice, beef, coffee solution, and apple slices)

- thermocouples to measure the temperature of the samples and the freezing medium
- a PC or equivalent to be used as a data logger (A minimum of six channels is necessary to simultaneously record the temperature of five samples and a freezing medium.)

6.4.2 Procedures

Before coming to the lab, the student should become familiar with typical calculations for obtaining the freezing point depression of a sugar solution, in addition to the freezing time for the products that will be frozen during the experiments. Therefore, theoretical values should be computed in advance. To do so, prepare four sugar and one sodium chloride solution. The following mixtures are suggested: 10% (w/w) NaCl, 30 and 50% sucrose, and 30 and 50% glucose. Approximately 200 ml of solution will be required.

(1) Locate five 250-ml volumetric flasks and label them.

(2) Weigh 25 g NaCl, 75 and 125 g sucrose, and 75 and 125 g glucose using weigh dishes.

(3) Transfer the weighed NaCl, sucrose, and glucose specimens into five prelabeled volumetric flasks.

(4) Add distilled (or deionized) water into the volumetric flasks until it reaches the designated marks.

(5) Cover the stoppers, holding them by hand while agitating the flasks by turning them up and down several minutes until all the solute dissolves.

(6) Locate twelve 250-ml plastic beakers and label them.

(7) Transfer the solution from the volumetric flasks into the beakers. The solutions from each flask should be distributed evenly into two beakers. Fill the last two beakers with distilled water.

(8) Divide the solutions in the 12 beakers into two groups: the first to be used for blast freezing, and the second for still air freezing.

(9) Put one group of solutions in the blast freezer and the other in the still air freezer (or cold room).

(10) Insert a thermocouple in each solution and start recording the temperature at 15-sec intervals. Also record the air temperature. Run the experiment until the theoretical freezing time for each solution is reached.

(11) Observe the samples and determine whether ice crystals have formed during the freezing process. Write down your observations.

(12) Proceed until all solutions are completely frozen.

(13) Compare the experimental frozen time with the theoretical frozen time obtained using the equations presented earlier in this chapter.

(14) Prepare the food samples to be frozen. The solid products may be cut in slabs of the same dimension as the model aluminum slab used for the experiments of convective heat transfer coefficient determination in Chapter 3. Liquid products may be poured into a can with the same dimensions as the model cylinder used. Freeze at least three products in the blast freezer and the same number of products in the cold room (still air).

(15) Insert a thermocouple in each sample and record the temperature at 1-min intervals. Make sure that the thermocouple tip is the geometrical center of sample. Proceed until the theoretical time has elapsed.

(16) Observe the samples and determine whether they are completely frozen. If not, continue the experiment until a completely frozen state is reached.

It is important to note the significance of determining whether a sample has reached a completely frozen state. Some analytical methods such as nuclear magnetic resonance (NMR) are used to assess the degree of freezing. However, for this lab, simple observation will be used to determine when the fully frozen state has been obtained.

6.4.3 Example 1

A comminuted meat product packaged in a box carton with dimensions 5 cm × 15 cm × 15 cm is frozen by direct contact between refrigerated plates that are −30°C with a convective heat transfer coefficient of 30 w/m² K. The heat conductivity is 1 w/m K. If the density of the meat is 950 kg/m³ and the initial temperature of the product is 15°C, compute the time required to freeze the product to −20°C.

6.4.3.1 SOLUTION

Using Nagaoka's equation requires knowledge of several product characteristics and related factors:

- The freezing point is −2°C (assumed).
- The specific heat of unfrozen water at 0°C(c_{pU}) is 3.52 kJ/kg K.
- The specific heat of ice (c_{pI}) is 2.05 kJ/kg K.

So

$$\Delta H' = [1 + 0.00445(288 - 271)][3.52(288 - 271) + 248.25 + 2.05(271 - 255)]$$

$$= 366.7 \text{ kJ/kg}$$

$$\beta_1 = \frac{0.15}{0.05} = 3$$

$$\beta_2 = \frac{0.15}{0.05} = 3$$

Then P and R can be determined from Figure 6.1:

$$P = 0.3 \text{ and } R = 0.085$$

Using Equation (6.3):

$$t_F = \frac{\rho \Delta H'}{T_F - T_\infty}\left(\frac{Pa}{h_c} + \frac{Ra^2}{k}\right)$$

$$= \frac{950(366.68)(1000)}{271-243}\left(\frac{0.3(0.05)}{30} + \frac{0.085(0.05)^2}{1}\right)$$

$$= 8864 \text{ sec}$$

$$= 2.46 \text{ hr}$$

6.4.4 Example 2

Joints of approximately spherical shaped meat with a maximum thickness of 18 cm are found to have a freezing time of 5 hr at a freezing medium temperature of –30°C. (a) How long will it take to freeze 15-cm-diameter joints in the same freezer? (b) Would it be worthwhile to increase the air velocity in the freezer to obtain faster freezing? (c) How much would freezing at –40°C reduce the freezing time?

6.4.4.1 SOLUTION

(1) If heat transfer to the freezing medium is mostly controlled by conduction within the product, Nagaoka's equation shows freezing time is proportional to a^2 because Pa/h is negligible. If controlled by convection, freezing time is proportional to a. In this example, the freezing time is large, and internal conduction will probably control the freezing time. Therefore, the freezing time for a 15-cm joint will be approximately 5 hr $\times (15/18)^2 =$ 3.5 hr. More precisely, a likely value for h is about 150 W/m³ K and k for meat is approximately 1 W/mK. Thus, $(Pa/h + Ra2/k)$ for the larger joint =

0.00165; for the smaller joint the term is $0.15/(24 + 1) = 0.0011875$. Because freezing time is proportional to this term, the smaller joint has a freezing time $= (0.0011875/0.00165) \times 5$ hr $= 3.6$ hr.

(2) Because Pa/h is much smaller than Ra^2/k, the effect of a change in air velocity (and hence h) on freezing time will be small. Doubling the velocity will increase h by a factor of about 1.74 (taking $h \propto \text{velocity}^{0.8}$). An increase of h from 100 to 174 will decrease $(pa/h + Ra^2/k)$ from 0.00165 to 0.0015224 and only reduce the freezing time by 7.7%.

(3) The freezing time is inversely proportional to $(T_F - T_\infty)$, so reducing the freezing medium temperature to $-40°C$ would reduce the freezing time to $5 \text{ hr} \times [-2-(-30)]/[-2-(-40)] = 3.68$ hr.

6.5 EXPECTED RESULTS

(1) Submit the freezing curves and plots of temperature versus time for all the products and solutions frozen in these experiments. Clearly show the initial freezing point.

(2) Compute the initial freezing point depression for the sugar solutions. How does this value agree with the experimental one?

(3) Compute the theoretical freezing time of the products used in this laboratory practice using both Planck and Nagaoka's equations. How do these results agree with the experimental data?

6.6 QUESTIONS

(1) What difference do you notice in predicting the freezing time using Nagaoka and Planck's equations? Which one agrees better with the experimental data? Why?

(2) Do you think that the predicting relations presented here for binary sugar solutions will hold for any kind of solution? What will be the major difficulty in trying to predict the freezing point of orange juice, for example?

(3) Does the rate of freezing have any effect on the freezing point or physical properties of the final product? Explain.

6.7 REFERENCES

1. Schwartzberg, H.G. 1990. Food freeze concentration. In: *Biotechnology and Food Process Engineering.* H.G. Schwartzberg and M.A. Rao, eds. Institute of Food Technologists, Chicago.
2. Heldman, D.R., and R.P. Singh. *Food Process Engineering.* 2 ed. AVI Publishers, Westport, CT.

CHAPTER 7

Drying of Foods: Part I. Tray Drying

7.1 INTRODUCTION

Drying foods is one of the most common methods of food preservation. The principle of drying is to decrease water availability for enzymatic reactions and microbial growth by removing free water from food products. Another objective of the drying process is the reduction of bulk volume and weight in the manufacture of convenient foods.

One must differentiate drying from evaporation, which is another process aimed at removing water from a food product. The difference between these two is that drying tends to remove most of the free water from a food system to decrease perishability. Evaporation, on the other hand, only removes part of the free water to reduce bulk volume prior to further the dehydration processes, as in the case of milk evaporation prior to spray drying.

During the drying process heat is required to remove water by vaporization of the liquid phase or direct sublimation from ice. This chapter is the first of three devoted to the study of three dehydration techniques: tray, spray, and freeze drying. In this section, the process of atmospheric drying in a tray or shelf dryer will be studied.

Tray drying is a popular technique because it does not require the use of highly specialized equipment as in spray and freeze drying. However, as in most dehydration methods, it is a combined heat and mass transfer operation that consists of exposing a food product to a heated air current that removes free moisture from the food. There are three major industrial methods of removing moisture from solid materials: (1) subjecting the material to a high-velocity stream of heated low-humidity air (air drying), (2) placing the material on a heated surface and waiting for the evaporation of moisture into the surrounding atmosphere (contact drying), and (3) subjecting the material to a low pressure and a heating source (vacuum drying).

63

The experiment described here is a batch-type operation in which food materials (i.e., apple slices or any other food materials) are evenly spaced on metal trays and placed in a drying chamber. Heated air is circulated over and parallel to the surface of the trays. Water loss in the product is monitored and recorded as weight loss in a digital balance weighing the trays connected to a data logger system.

7.2 THEORY AND CALCULATIONS

Drying usually occurs in a number of stages, each characterized by a different dehydration rate. Figures 7.1 and 7.2 show typical drying characteristics of wet solids in air [1]. If one ignores the very early stage during which the product temperature is rising, there are two major stages of drying: (1) the constant rate period in which water is evaporated from saturated surfaces (making the evaporation rate to the moving air stream the limiting factor of the dehydration process) and (2) the falling rate period in which moisture is bound or held within the solid matrix. As the saturated surface becomes smaller, water has to overcome a resistance to flow through the normally porous food material. The moisture content at which the rate of drying suddenly decreases (or the point at which the rate begins to fall) becomes the critical moisture content of the process. The minimum moisture the product can reach when exposed to a specific external medium (e.g., 60°C air with 30% *RH*) after a theoretical infinite time is called the equilibrium moisture content.

7.2.1 Constant Drying Rate Period

If heat transfer is solely by convection from air drying, the surface temperature drops to the wet bulb temperature of the air as the latent heat of evaporation is taken up by the film of water at the surface. The wet bulb temperature can be determined from psychrometric data for moist air and the drying rate remains constant as long as free water exists at the surface.

If L is the depth of the innermost section of the material from the drying surface and similarily if drying occurs from both sides, then L is half the total thickness of the solid. ρ_s is the dry matter density of the material (kg dry matter/m³ of the material), M_s is the mass of dry solids, and A is the area of the top surface of the solid, then the volume V of material is:

$$V = (\text{surface area}) * (\text{depth}) = (A) * (L) = M_s / \rho_s \qquad (7.1)$$

Rearranging Equation (7.1) results in:

$$\frac{A}{M_s} = \frac{1}{L\rho_s} \qquad (7.2)$$

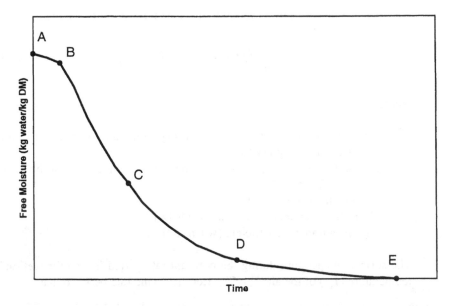

FIGURE 7.1. Typical drying curve of loss of moisture with time for a solid (reprinted from Reference [1], p. 266, with permission of Chapman & Hall).

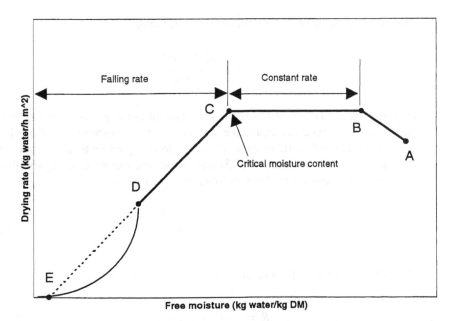

FIGURE 7.2. Drying rate curve (reprinted from Reference [1], p. 267, with permission of Chapman & Hall).

Heat balance: latent heat of evaporation = heat transferred to the medium

$$-\frac{dw}{dt} \cdot \left(M_s\right)\left(h_{fg}\right) = hA\left(T_a - T_s\right) \tag{7.3}$$

where

$-dw/dt$ = loss of moisture per unit of dry solid and per unit of time [(kg H_2O)/(kg DM) s]

h_{fg} = latent heat of vaporization at the surface temperature of the material [J/(kg H_2O)]

T_a = air temperature [K]

T_s = surface temperature of the solids [K]

h = heat transfer coefficient [w/(m² K)]

The surface temperature during the constant rate period is also the wet bulb temperature (T_{wb}) of the air: $T_s = T_{wb}$. Let the drying rate be R_c so that:

$$R_c = -\frac{dw}{dt} = \frac{h\left(T_a - T_s\right)}{h_{fg}} \cdot \frac{A}{M_s} \tag{7.4}$$

Substituting Equation (7.2) in Equation (7.4) yields:

$$R_c = \frac{h\left(T_a - T_s\right)}{h_{fg} L\rho_s} \tag{7.5}$$

Equation (7.5) can thus be used to calculate the constant rate of drying from the heat transfer coefficient and the wet and dry bulb temperatures of the drying air for a bed of particles with drying air flowing parallel to the surface. An expression similar to Equation (7.5) may be derived using the same procedure as above for cubes with sides L evaporating water on all sides:

$$R_c = \frac{6h\left(T_a - T_s\right)}{h_{fg} L\rho_s} \tag{7.5a}$$

For a brick-shaped solid with sides a and 2a and thickness L:

$$R_c = \frac{h\left(T_a - T_s\right)}{h_{fg} L\rho_s}\left[\frac{3}{a} + \frac{2}{L}\right] \tag{7.5b}$$

For solid materials in slab form on trays or belts in a dryer, the air flow is commonly either parallel or perpendicular to the slab surface. The heat transfer coefficient in these two cases can be calculated [3] as follows:

(1) For parallel air flow:

$$h = 14.305 \, G^{0.8} \tag{7.6}$$

(2) For perpendicular air flow:

$$h = 413.5 \, G^{0.37} \tag{7.7}$$

where G is the air mass velocity in kg/m²s.

7.2.2 Materials with One Falling Rate Stage Where the Rate of the Drying Curve Goes through the Origin

The drying rate curve for these materials shows a constant rate (R_c) from the initial moisture content (w_0) to the critical moisture (w_c). The drying rate then falls in a linear relationship with the decreasing moisture content until it becomes 0 at $w = 0$. The drying times required to reach a w moisture content in either of the drying stages are:

$$\text{Constant rate: } R_c = -\frac{dw}{dt}$$

The total time (t_c) for the constant rate stage is:

$$t_c = \frac{w_0 - w_c}{R_c} \tag{7.8}$$

At the falling rate period:

$$-\frac{dw}{dt} = \frac{R_c}{w_c} \cdot w \tag{7.9}$$

Rearranging and integrating Equation (7.9):

$$\int_{t_c}^{t} dt = \frac{w_c}{R_c} \int_{w_c}^{w} \frac{dw}{w} \tag{7.10a}$$

$$t - t_c = \frac{w_c}{R_c} \ln\left(\frac{w_c}{w}\right) \tag{7.10b}$$

The total time from w_0 to w in the falling rate stage can be calculated by transferring t_c in Equation (7.8) into Equation (7.9):

$$t = \frac{w_0 - w_c}{R_c} + \frac{w_c}{R_c} \ln\left(\frac{w_c}{w}\right) \qquad (7.11)$$

Equation (7.11) shows that if a material exhibits only the falling rate stage of drying and the drying rate is 0 only at $w = 0$, the drying time required to reach a desired moisture content can be determined from the constant rate (R_c) and critical moisture content (w_c). For these materials, w_c is usually the moisture content when water activity (a_w) starts to drop below 1 in a desorption isotherm.

7.2.3 Materials with More than One Falling Rate Stage

Most solid food exhibits the drying behavior exhibited in Figures 7.1 and 7.2, and the drying time in the constant rate follows Equation (7.8). However, because the rate of drying against the moisture content plot no longer goes to the origin from the point w_c, R_c, Equation (7.9) cannot be used for the falling rate stage. If the rate versus the moisture content line is extended to the abscissa, the moisture content that has a rate of 0 may be designed as the residual moisture content (w_r). For the first falling period:

$$\frac{d(w - w_{r1})}{dt} = \frac{R_c}{w_{c1} - w_{r1}}(w - w_{r1}) \qquad (7.12)$$

Integrating Equation (7.12) and using Equation (7.8) for drying time in the constant rate zone, drying to a moisture content w in the first falling rate stage takes:

$$t = \frac{w_0 - w_{c1}}{R_c} + \frac{w_{c1} - w_{r1}}{R_c} \ln\left(\frac{w_{c1} - w_{r1}}{w - w_{r1}}\right) \qquad (7.13)$$

where w_{c1} and w_{r1} represent the critical and residual moisture contents for the first falling rate stage of drying, respectively. The time required for a w drying moisture content in the second falling rate stage is:

$$t = \frac{w_0 - w_{c1}}{R_c} + \frac{w_{c1} - w_{r1}}{R_c} \cdot \ln\left(\frac{w_{c1} - w_{r1}}{w_{c2} - w_{r1}}\right)$$

$$+ \frac{w_{c1} - w_{r1}}{R_c} \cdot \frac{w_{c2} - w_{r2}}{w_{r2} - w_{r1}} \cdot \ln\left(\frac{w_{c2} - w_{r2}}{w - w_{r2}}\right) \qquad (7.14)$$

The drying process clearly shows that parameters such as constant and falling rate periods and critical and equilibrium moisture contents characterize the process. A plot of drying rate versus time is a common way to analyze the drying data. The influence of external variables such as air temperature and velocity will be studied during the experiments.

A typical drying curve of a food material is shown in Figure 7.1. The rate of drying (drying flux) has been defined in several ways in the literature, but for the purpose of this experiment it is obtained from:

$$N_c = -\frac{M_s}{A}\frac{dw}{dt} \tag{7.15}$$

where

N_c = drying flux (kg H_2O m^{-2} h^{-1})
M_s = total solids in the sample (kg)
A = area exposed to air flow (m^2)
$-dw/dt$ = moisture loss per unit of time (kg H_2O) (kg dry matter)$^{-1}$ h^{-1}

To obtain a drying curve, free moisture should be plotted against time. Notice that the free moisture is expressed in kg water/kg dry solids, where the initial value is obtained from the moisture determination test done in a vacuum oven, and the other values from the data of sample weight. Remember the weight loss of the product is due only to water evaporation.

To obtain the drying rate curve, plot the drying rate (N_c) against free moisture and use Equation (7.15) to obtain the rate of drying. The values M_s and A are readily available and constant throughout the process. To compute $-dw/dt$, assume a constant rate of drying within each interval, and find the loss of water per unit of time.

7.3 OBJECTIVES

(1) To obtain a typical drying curve of a food product by identifying the constant and falling rate periods and the critical and equilibrium moisture contents
(2) To understand the influence of external parameters such as drying air flow rate and temperature, on the drying characteristics of a product

7.4 MATERIALS AND METHODS

7.4.1 Materials

The following materials are required for the experiments:

- tray dryer with adjustable air flow and temperature [The schematic diagram of a tray dryer is shown in Figure 7.3. An example is an Armfield Tray Dryer (model UOP-8, Ringwood Hampshire, England) 3 kW with a maximum air velocity through the outlet duct of 1.5 m/sec, maximum temperature 80°C (lower at higher air flows; see the maximum temperature vs. air flow figure in the operation manual for the equipment), maximum wet material capacity 4 kg, two psychometric measurement ports (before and after the material trays), and four trays with approximate dimensions of 27cm × 17.5 cm for holding the material while drying.]
- apple slices or other raw material for drying
- 4-kg digital balance with 1-g resolution mounted in the dryer for continuous weighing
- aspirating psychrometer with a resolution of 1°C
- digital anemometer to measure the air flow in the dryer
- PC or equivalent with appropriate interface card for data acquisition (The computer will be connected to a balance to automatically record the weight of the material being dried.)
- precision balance with at least 1-mg resolution to be used for weighing the sample material before and after vacuum drying
- vacuum oven for moisture determination

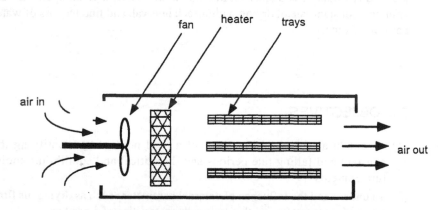

FIGURE 7.3. Schematic diagram of a tray dryer.

7.4.2 Procedures

(1) Grouping: Because each experiment may take 3 hrs or more, the class should be divided into several groups. The data should then be shared within groups, with every student expected to conduct data analyses of each experiment.

(2) Because the tray dryer is fully automatic, the computer will record the data; however, the students are expected to read the manual and be familiar with the equipment before conducting the experiments.

(3) When starting the dryer, preheat to the desired conditions (e.g., 70°C). Be sure that the trays are installed so that air flow restriction will be approximately the same as during the drying process. Next select the first air flow rate (1.5 m/sec). A graph of air temperature against flow at maximum power settings can be found in the operational instructions for the Armfield Tray Dryer. For the first experiment, it is advisable to select an air flow of about 50% full range and an air dry bulb temperature of about 70°C.

(4) Check the water level in the aspirating psychrometer, and if necessary, fill with deionized water. Record the wet and dry bulb temperature of the air and the air flow using the digital anemometer.

(5) Cut the apple in uniform slices, trimming them into a well-defined shape so that the area of each specimen can be computed. Save some samples of the raw material for further moisture determination.

(6) Evenly distribute the food product to be dried among the trays, and record the initial weight.

(7) Take measurements of the product weight at 30-sec intervals during the first 5 min and at 1-min intervals thereafter. Next, measure the wet and dry bulb temperature of the air, which should remain constant during the drying operation if there is good air circulation in the lab and no significant change in room conditions. The experiment should run until the equilibrium moisture content is reached, which will occur after observations of almost no change in the product weight.

(8) When the above steps are completed, take a sample of dry material for moisture determination by weighing samples of the raw and dried material and placing them in the vacuum oven at 70°C and an absolute pressure not higher than 100 mm Hg for at least 24 hr. Weigh the samples again after drying, and compute the moisture content as the difference in product weight. Express the moisture in dry basis (db), which is kg of water/kg dry solids.

(9) For the other experiments, use dry bulb temperatures of 60 or 80°C with an air flow rate of 75 or 100% of equipment capacity.

7.5 EXAMPLES

The following examples demonstrate the typical calculations in the drying process. Note that the calculation of drying time is based on the experimental drying data (i.e., as in the drying rate curve).

7.5.1 Example 1

Figure 7.4 shows the drying curve for apple slices blanched in 10% sucrose solution and dried in a cabinet dryer using air in parallel flow at a velocity of 3.65 m/sec; $T_{db} = 170°F$ (76.7°C) and $T_{wb} = 100°F$ (37.8°C) for the first 40 min, and $T_{db} = 160°F$ (71.1°C) and $T_{wb} = 110°F$ (43.3°C) for the rest of the drying period. The initial moisture content was 85.4% (wet basis), and the slices were in a layer 0.5 in. (0.0127 m) thick. Calculate the drying time needed to reach a moisture content of 13% (wet basis).

7.5.1.1 SOLUTION

The drying rate curve was obtained by drawing tangents to the drying curve at the designated moisture contents and determining the slopes of the tangents as shown in Figure 7.5. Because two breaks in the drying rate curve can be

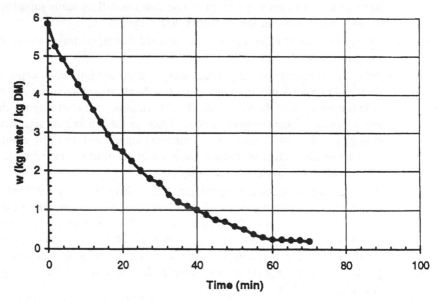

FIGURE 7.4. Drying curve and drying rate as a function of the moisture content of apple slices.

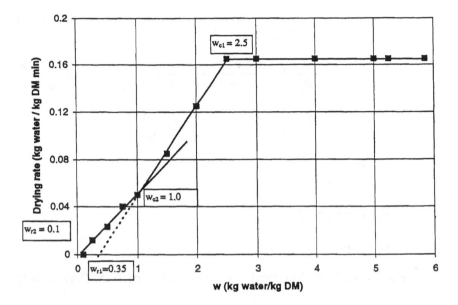

FIGURE 7.5. The drying rate curve of apple slices, showing several breaks in drying rate.

observed, it is clear that it is a material with more than one falling rate stage. To calculate the drying time, Equation (7.14) may be used. [It is good practice to list all the parameters required in Equation (7.14) and convert them to a constant units system such as SI or the American Engineering System]:

$$w_0 = \frac{85.4}{100 - 85.4} = 5.85 \left(kg\ H_2O / kg\ DM \right)$$

$$w = \frac{13}{100 - 13} = 0.149 \left(kg\ H_2O / kg\ DM \right)$$

where *DM* is dry matter, and the following parameters can be read from the drying curve and drying rate curve:

$w_{c1} = 2.5$ (kg H$_2$O/kg *DM*)
$w_{c2} = 1.0$ (kg H$_2$O/kg *DM*)
$w_{r1} = 0.25$ (kg H$_2$O/kg *DM*)
$w_{r2} = 0.10$ (kg H$_2$O/kg *DM*)
$R_c = 0.165$ (kg H$_2$O/min kg *DM*)

The drying time required to obtain a moisture content of 13% (wet basis) can then be calculated using Equation (7.14):

$$t = \frac{w_0 - w_{c1}}{R_c} + \frac{w_{c1} - w_{r1}}{R_c} \ln \frac{w_{c1} - w_{r1}}{w_{c2} - w_{r1}} + \frac{w_{c1} - w_{r1}}{R_c} \cdot \frac{w_{c2} - w_{r2}}{w_{c2} - w_{r1}} \ln \frac{w_{c2} - w_{r2}}{w - w_{r2}}$$

$$= \frac{5.85 - 2.5}{0.165} + \frac{2.5 - 0.35}{0.165} \ln \frac{2.5 - 0.35}{1.0 - 0.35} + \frac{2.5 - 0.35}{0.165}$$

$$\cdot \frac{1.0 - 0.1}{1.0 - 0.35} \ln \frac{1.0 - 0.1}{0.15 - 0.1}$$

$$= 20.3 + 15.6 + 52.5$$

$$= 88.4 \text{ min}$$

7.5.2 Example 2: Graphical Integration in Falling Rate Drying Period

A batch of wet solid whose drying rate curve is represented by Figure 7.6 is to be dried from a free moisture content of $w_0 = 0.38$ kg H_2O/kg dry solid to $w_2 = 0.04$ kg H_2O/kg dry solid. The weight of the dry solid is $M_s = 399$ kg and $A = 18.58$ m² of top drying surface. Calculate the time for drying, noting that $M_s/A = 399/18.58 = 21.5$ kg/m².

7.5.2.1 SOLUTION

From Figure 7.6(b), the critical free moisture content is $w_c = 0.195$ kg H_2O/kg dry solid. Hence, drying occurs during the constant rate and falling rate periods. For the constant rate period, $w_0 = 0.38$ and $w_1 = w_c = 0.195$. From Figure 7.6(b), $R_c = 1.51$ kg H_2O/h m² or:

$$R_c = 1.51 \frac{\text{kg}(H_2O)}{\text{k} \cdot \text{m}^2} \cdot \frac{18.58(\text{m}^2)}{399(\text{kgDM})} - 0.0703 \text{ kg } H_2O/\text{h kg } DM$$

Substituting into Equation (7.8):

$$t_c = \frac{w_0 - w_c}{R_c}$$

$$= \frac{0.38 - 0.195}{0.073}$$

$$= 2.53 \text{ h}$$

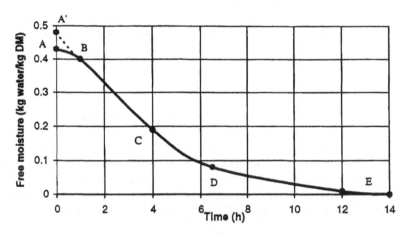

(a) Plot of data as free moisture versus time

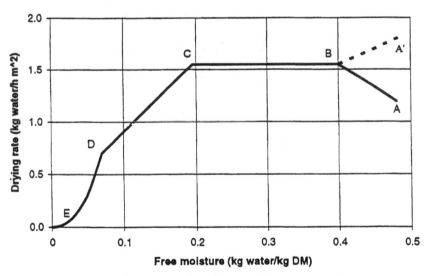

(b) Rate of drying curve as rate versus free moisture content

FIGURE 7.6. The drying curve of wet granules.

For the falling rate period, the rate of drying R is not constant but decreases when drying proceeds past the critical free moisture content (w_c). When the free moisture content is 0, so is the rate. The time for drying any region between w_1 and w_3 can be computed by the following expression:

$$t = \int_{w_3}^{w_1} \frac{dw}{R}$$

(7.16)

If the rate is constant, Equation (7.16) can be integrated to produce Equation (7.8). However, in the falling rate period, R varies. For any shape of falling rate drying curve, Equation (7.16) can be graphically integrated by plotting $1/R$ versus w and determining the area under the curve. Reading values of R for various values of w from Figure 7.6(b), Table 7.1 is prepared.

A plot of $1/R$ versus w is made and shown in Figure 7.7. The area under the curve from $w_1 = 0.195$ (point C) to $w_3 = 0.040$ is determined by:

$$\text{Total area} = A_1 + A_2 + A_3 + A_4 + A_5$$

$$= (0.05 - 0.04)*(79.54 + 58.04)/2$$

$$= (0.065 - 0.05)*(58.04 + 30.24)/2$$

$$= (0.1 - 0.065)*(30.24 + 23.86)/2$$

$$= (0.15 - 0.10)*(23.86 + 17.75)/2$$

$$= (0.195 - 0.150)*(17.75 + 14.22)/2$$

$$= 0.6879 + 0.6621 + 0.9468 + 1.040 + 0.7193$$

$$= 4.06\ h$$

TABLE 7.1. for Example 2

w (kg H_2O/kg DM)	R (kg H_2O/h kg DM)	1/R (kg H_2O/h kg DM)
0.195	0.073	14.22
0.150	0.0563	17.75
0.100	0.0419	23.86
0.065	0.033	30.24
0.050	0.0172	58.04
0.040	0.0126	79.54

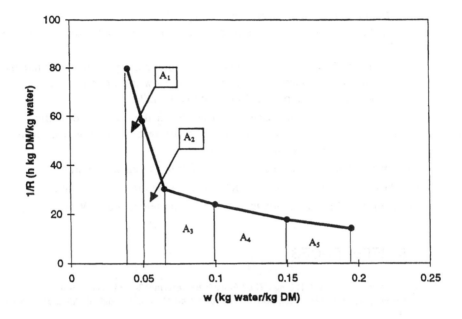

FIGURE 7.7. Graphical integration for falling rate period in Example 2.

Substituting into Equation (7.16) to obtain the drying time in the falling rate period:

$$t = \int_{w_3}^{w_1} \frac{dw}{R} = \text{total area under curve in Figure 7.7}$$

$$= 4.06 \, h$$

Total time is: $2.53 + 4.06 = 6.59h$

7.6 EXPECTED RESULTS

(1) Submit the drying curve and drying rate curves for each of the experiments and clearly identify the constant rate and falling rate periods and the critical and equilibrium moisture contents.

7.7 QUESTIONS/INSTRUCTIONS

(1) Why was it necessary to have uniform product slices prior to drying?

(2) Would you recommend that the drying cabinet be preheated before load-ing the food product, or that the product be inserted first before turning on the heat?

(3) For one of the experiments, compute a theoretical drying time to obtain the final moisture content reached in the lab. How does this time agree with the actual time you obtained in the experiment?

(4) Does the air velocity have an influence on the constant or falling rate period and/or critical or equilibrium moisture content? Suggest physical explanations for the observed results.

(5) Answer the previous question with reference to air temperature.

(6) Answer Question 4 with respect to the drying air's relative humidity.

(7) Is there a practical limitation for the drying temperature? Why?

7.8 REFERENCES

1. Heldman, D. R. and R. P. Singh. *Food Process Engineering.* 2nd ed. AVI, Westport, CT.
2. Geankoplis, C. J. 1983. *Transport Process and Unit Operations.* 2nd ed. Allyn and Bacon, Boston.

CHAPTER 8

Drying of Foods: Part II. Spray Drying

8.1 INTRODUCTION

Spray drying is a unit operation widely used in the food processing industry. Basically, a spray dryer system has five essential elements as shown in Figure 8.1: an air heater, drying chamber, system for dispersing material to be dried as droplets in the drying chamber, system for collecting dry particles from the air, and one or more blowers for moving air through the system.

Air may be heated by direct firing of gas or fuel in an airstream or indirectly through a heat exchanger. Indirect heating is often used in the food industry to avoid contamination of food with carbon or dirt from fuel and contact with combustible products that could impart foreign flavors or odors.

The primary function of the drying chamber is to provide an intimate mixing of hot air with finely dispersed droplets of the material to be dried in such a way that drying proceeds adequately and that the dry particles have the desired characteristics. The drying chamber may be in the form of a horizontal box or tall vertical tower and may have a relatively simple or extremely complex flow pattern for both air and feed droplets.

The primary function of atomization is to generate small droplets that create a large surface area for moisture evaporation. In addition, an atomizer acts as a metering device to control the flow rate of product into the dryer. The three types of extensively used atomizers are centrifugal, pressure spray head, and fluid nozzles.

Coffee, eggs, milk, soups, and baby foods are among the foodstuffs normally spray-dried. The feed may be a solution, suspension, or paste, but the final product is normally a powder with variable physical properties according

79

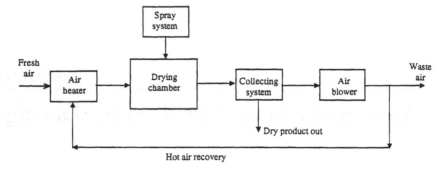

FIGURE 8.1 Elements of a spray dryer system.

to the design and operational characteristics of the spray dryer. The process begins with a slurry containing 12 to 70% solids, depending on the specific application and feed material.

The advantages of this operation are that the process is continuous and adaptable to full automatic control, which increases product output without adding labor. However, the process is quite specific because spray dryers are not general purpose types of equipment. Their narrow range of products serves to improve the efficiency and quality of the final product.

The purpose of this laboratory exercise is to familiarize the student with a spray-drying operation and some of the variables that affect the characteristics of the final product. Determination of heat and mass balance will be practiced to find the thermal and evaporative efficiency of the system.

8.2 OBJECTIVES

(1) To become familiar with a spray-drying operation
(2) To calculate the overall efficiency of a spray dryer
(3) To measure the moisture content of the final product at several air outlet temperatures
(4) To observe differences in the final product using different total solids (TS) in the feed
(5) To observe how the quality of the final powder is affected by the temperature of the air, feed, and outlet; solids in the feed; and degree of optimization

8.3 MATERIALS AND METHODS

8.3.1 Apparatus and Materials

The following materials are required to do the experiments in this chapter:

- spray dryer (ANHYDRO™ model or equivalent)
- digital anemometer to measure the flow rate through the spray dryer
- psychrometer and thermometer to measure the inlet and outlet temperature and relative humidity
- temperature-controlled vacuum oven, electronic balance with a precision of ±0.001 grams, and metal sample holders to determine the moisture of the final product and total solids of the feed
- particle size analyzer to obtain the particle size distribution of the final powder (standard sieves from 12 to 100 mesh will be used)
- magnetic stirrer and heater and distilled water to rehydrate the powder obtained from the spray dryer
- 50-ml graduate cylinders
- skim or whole milk with at least three different total solids content (TS) to observe how the composition of the raw material affects the final product

8.3.2 Procedure

(1) Become familiar with the operation of the dryer by reading the instruction manual. Warm up the dryer and begin the operation with water. In the case of the ANHYDRO™ model, the centrifugal atomizer should be used, which will require waiting until it reaches steady state and the air outlet temperature is about 90°C.

(2) Change to the selected product (e.g., whole milk) and adjust the flow to keep the same air outlet temperature. When steady conditions are reached, measure the airflow leaving the dryer with the anemometer. (It is better to measure the flow at the outlet because the inlet flow has a conical shape that may yield an erroneous reading.) Also measure the inlet and outlet temperature and relative humidity with the psychrometer.

(3) Measure the temperature of the outlet product and take a sample for later moisture determination by weighing approximately 10 grams of product, placing it in the vacuum oven, setting the temperature to 90°C, and then pulling the vacuum. The absolute pressure inside the oven must not exceed 100 mm Hg when the samples are in the oven for the required 24 hr. After vacuum drying, weigh the samples again and compute the moisture content as a percentage of initial weight (wet basis).

(4) Study the effect of the centrifugal atomizer speed (i.e., using 2000, 4000, and 6000 rpm). Using the sifter, determine the particle size distribution in each case by screening the powder through a series of sieves and determining the percentage of weight retained on each sieve. (A frequency histogram of the particle size may be drawn to show how the speed of the rotating wheel affects the mean droplet diameter.)

(5) Repeat the experiment with an exhaust air temperature of 80 and 100°C at a fixed rpm of the rotating wheel. (Notice that you can obtain the same result by increasing the inlet air temperature or decreasing the rate of product entering the dryer.) Try two different combinations to obtain the desired exhaust temperature.

(6) Repeat three of the experiments described above using the two fluid atomizing nozzles. Use 5, 9, and 12 psi for your three levels of air pressure, and compare the product obtained each time.

(7) Select one operating condition for the dryer and repeat the experiment using milk with a different *TS*.

(8) Rehydrate the product obtained in each of the experiments. [The normal proportion used for home consumption is 10.5% for skim milk and 12.5% for whole milk (wb).] Prepare the dilution according to the product dried in the lab using distilled water at 25°C, and report on the flavor and appearance of the observed suspension.

8.3.3 Calculations

Moisture balance in the spray dryer:

- moisture entering in feed: $M_s(W_s)_1$
- moisture entering in hot air: $G_a H_1$
- moisture leaving the dryer in the dried product: $M_s(W_s)_2$
- moisture leaving in the exhaust drying air: $G_a H_2$

where

M_s = dry solid of feed (kg dry solid)
$(W_s)_1$ = water content of feed (kg water/kg dry solid)
$(W_s)_2$ = water content of the product (kg water/kg dry solid)
G_a = dry air of the air entering the drying chamber (kg dry air)
H_1 = absolute humidity of air entering the drying chamber (kg water/ kg dry air)
H_2 = absolute humidity of air leaving the drying chamber (kg water/ kg dry air)

Under steady-state operation, there is no accumulation in the chamber. The following equation holds for mass balance in the drying chamber:

$$input = output$$

Thus:

$$M_s(W_s)_1 + G_a(H_1) = M_s(W_s)_2 + G_a H_2 \qquad (8.1)$$

or

$$M_s\left[(W_s)_1 - (W_s)_2\right] = G_a(H_2 - H_1) \qquad (8.2)$$

The enthalpy or heat balance in the spray dryer is as follows:

- enthalpy of air entering dryer: $G_a(Q_a)_1$
- enthalpy of feed entering dryer: $M_s(Q_s)_1$
- enthalpy of exhaust drying air: $G_a(Q_a)_2$
- enthalpy of dried solid: $M_s(Q_s)_2$

$$heat\ input = heat\ outlet + heat\ loss$$

Thus

$$G_a(Q_a)_1 + M_s(Q_s)_1 = G_a(Q_a)_2 + M_s(Q_s)_2 + Q_L \qquad (8.3)$$

where

$$Q_L = heat\ loss\ from\ the\ dryer$$

Notice that the heat loss can be very small if one has a well-insulated chamber though there are some cases of defective insulation or special chamber design in which air is cooling the walls to allow handling of special products.

The enthalpy of feed $(QS)_1$ is the sum of the enthalpy of the dried solid and the moisture as a liquid.

Thus:

$$(Q_s)_1 = C_{DS}(\Delta T) + (W_s)_{1CW} \Delta T \qquad (8.4)$$

where

C_{DS} = heat capacity of dry solid
C_W = heat capacity of moisture (in liquid form)
ΔT = difference between feed temperature and a reference point (normally chosen as 0°C)

The enthalpy of the drying air (Q_a) can be computed either at the inlet or outlet of the drying chamber by:

$$Q_a = C_S(\Delta T) + H\lambda \qquad (8.5)$$

where

C_S = humid heat 1.005 + 1.88 H (SI units) or 0.24 + 0.45 H (English units)
λ = latent heat of vaporization
H = humidity

The overall thermal efficiency $\eta_{overall}$ is defined as the fraction of the total heat supplied to the dryer used in the evaporation process. In the case of a truly adiabatic system, it can be approximated to the relation:

$$\eta_{overall} = \left[\frac{T_1 - T_2}{T_1 - T_0} \right] \cdot 100 \qquad (8.6)$$

where

T_1 = temperature of the air entering the spray dryer
T_2 = temperature of air leaving the spray dryer
T_0 = temperature of the atmospheric air

Evaporative efficiency is defined as the ratio of the actual evaporation capacity to the capacity obtained in the ideal case of air leaving the dryer at saturation conditions. It can be approximated to the relationship:

$$\eta_{evap} = \left[\frac{T_1 - T_2}{T_1 - T_{sat}} \right] \cdot 100 \qquad (8.7)$$

where

T_{sat} = adiabatic saturation temperature corresponding to T_1

The residence time of the product is an important parameter in the design and selection of operating conditions for the dryer because it allows moisture to leave the product. The minimum residence time can be computed by dividing the volume of the chamber by the air volumetric flow (computed at the average chamber temperature). In the case of a cylindrical drying chamber with a conical base (as in the ANHYDRO™ spray dryer), the volume can be computed by the following assuming a 60° cone angle:

$$V = 0.7854 D_{ch}^2 \left(h' + 0.2886 D_{ch} \right) \tag{8.8}$$

where

    h' = cylindrical height
    D_{ch} = diameter of the drying chamber

Notice that the residence time computed this way is a minimum time: most of the particles will stay in the chamber longer due to recirculating air currents inside the dryer, product adherence to the walls, and air flowing at a lower than average velocity [1].

8.3.4 Drying Times in Spray Drying

The removal of moisture from droplets using spray drying to obtain food solids is accomplished by simultaneous heat and mass transfer. Spray drying differs from other dehydration methods in that a large percentage of moisture is removed during a constant rate drying period when the moisture content is very high. During this period, evaporation of moisture from the droplet takes place in the same manner as for a pure water droplet. Simultaneous heat and mass transfer occurs as (1) heat for evaporation is transferred by conduction and convection from the heated air to the droplet surface and (b) the vapor produced by moisture evaporation is transferred by diffusion and convection from the droplet surface to the heated air. The rate at which the process occurs is a function of factors such as air temperature, humidity or vapor pressure, transport properties of air, droplet diameter and temperature, relative velocity, and the nature of the solid in the liquid droplet.

Partial differential equations are required to accurately describe the heat and mass transfer occurring around a liquid droplet during evaporation. The analytical solution to these equations is difficult, but the following equations provide an approximation for the drying time in constant and falling rate periods:

$$t_c = \frac{\rho_L \lambda d_0^2}{8 k_g \left(T_a - T_w \right)} \tag{8.9}$$

$$T_j = \frac{\rho_p d_c \lambda (w_c - w_e)}{6h\Delta T_{ave}}$$ (8.10)

The total drying time for a droplet in static air ($N_{Rc} \approx 0$) would be the following:

$$t = \frac{\rho_L \lambda d_0^2}{8k_g(T_a - T_w)} + \frac{\rho_p d_c \lambda (w_c - w_e)}{6h\Delta T_{ave}}$$ (8.11)

8.3.5 Example

In this laboratory exercise, the main objective is to gain practical experience in spray dryer operation and an understanding of the effect of factors such as air temperature and degree of atomization on final powder quality. The following example illustrates a concept for spray drying where a cocurrent spray dryer is being planned for drying a product with 25% total solids. Ambient air is assumed at 25°C and 60% relative humidity. The maximum temperature of the air entering the dryer is 100°C due to the product's heat sensitivity. The product droplets generated by the atomizer have an estimated range in diameters from 40 to 65 μm and their critical moisture content is 48% (wet basis). The product leaves the dryer in equilibrium with an air temperature of 50°C and a final moisture content of 5% (wet basis). Assuming the latent heat of vaporization is 2360 kJ/kg, estimate the size of the drying chamber required to dry the product when the feed rate is 20 kg/min at 15°C.

8.3.5.1 SOLUTION

(1) In this design problem, the two important factors determining the size of the drying chamber are volume of air required and drying time. These can be found by mass and energy balance.

(2) Mass balance for the product:

$$20 = m_p + m_w^0 \text{ [kg/min]}$$ (8.12)

where m_p is the production rate (kg/min), and m_w^0 is the removed water from the feed (kg water/min).

Mass balance for the drying solid in the product:

$$20 (0.25) = m_p (0.95) \text{ [kg/min]}$$ (8.13)

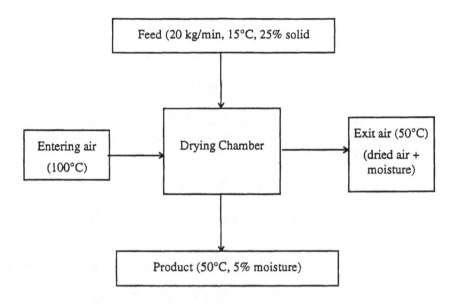

Solving for m_p from Equation (8.13):

$$m_p = 5.263 \text{ kg/min}$$

substituting $m_p = 5.263$ kg/min into Equation (8.12) to find m_w^0:

$$m_w^0 = 20 - 5.263 = 14.737 \text{ kg water/min}$$

(3) Energy balance on the drying chamber:

At steady state and assuming the heat loss to the environment is very small and can be neglected, $E_{in} = E_{out}$

or

$$E_f + E_a^i = E_p + E_a^0 + R_{evp} \qquad (8.14)$$

where E_f, E_a^i, E_p, E_a^0, and E_{evp} are enthalpy of feed, entering air, product, exiting air, and latent heat of water evaporated from the feed, respectively. These values can be calculated as follows:

$$E_f = c_f(20)(288 - 273) \quad [\text{kj}/\text{min}] \qquad (8.15a)$$

$$E_a^i = c_a^i\left(m_a^i\right)(373 - 273) \quad [\text{kj}/\text{min}] \qquad (8.15b)$$

$$E_p = c_p \, (m_p) \, (323 - 273) \quad \text{[kJ/min]} \tag{8.15c}$$

$$E_a^0 = c_a^0(m_a^0) \, (323 - 273) \quad \text{[kJ/min]} \tag{8.15d}$$

$$E_{evp} = \lambda \, (m_w^0) \, \text{[kJ/min]} \tag{8.15e}$$

The energy balance uses 0°C (273 K) as a reference for energy content, and the specific heat can be estimated from the moisture content indicated in Siebel's model [4]:

$$c_p = 0.837 + (3.349)w$$

where w is water content, and the specific heat for feed and product is estimated as follows:

$$c_f = 0.837 + (3.349) \, (0.75) = 3.349 \text{ kJ/kg} \tag{8.16a}$$

$$c_p = 0.837 + (3.349) \, (0.05) = 1.004 \text{ kJ/kg} \tag{8.16b}$$

The specific heat of entering and exiting air is estimated by the following equation [5]:

$$c_p = 1.005 + 1.88H \text{ [kJ/kg]} \tag{8.17}$$

where H is the absolute humidity of the air (kg water/kg dry air). The absolute humidity of entering and exiting air can be read from a psychrometric chart:

$H^i = 0.013$ kg water/kg dry air
$H^0 = 0.032$ kg water/kg dry air

So the specific heat of entering and exiting air is:

$$c_a^i = 1.005 + (1.88) \, (0.013) = 1.029 \text{ kJ/kg} \tag{8.18a}$$

$$c_a^0 = 1.005 + (1.88) \, (0.032) = 1.063 \text{ kJ/kg} \tag{8.18b}$$

Substituting Equations (8.16) and (8.18) into Equation (8.15):

$$E_f = (3.349) \, (20) \, (288 - 273) = 1,004.7 \text{ kJ/min} \tag{8.19a}$$

$$E_a^i = (1.029)(m_a^i)(373 - 273) = (1.029) \, m_a^i \text{ kJ/min} \tag{8.19b}$$

$$E_p = (1.004) \, (5.263) \, (323 - 273) = 264.2 \text{ kJ/min} \tag{8.19c}$$

$$E_a^0 = (1.063)\left(m_a^0\right)(323 - 273) = (53.15)m_a^0 \text{ kj/min} \qquad (8.19d)$$

$$E_{evp} = \lambda\left(m_u^0\right) = (2360)(14.737) = 34,779.32 \text{ kg/min} \qquad (8.19e)$$

Substituting Equation (8.17) into Equation (8.14) and considering that $m_a^i = m_a^0$ because inlet drying air is equal to outlet drying air,

$$1004.7 + (102.9)m_a^i = 264.2 + (53.15)m_a^0 + 34,779.32 \text{ [kJ/min]} \quad (8.20)$$

Solving for m_a^i:

$$m_a^i = 684.19 \text{ kg dry air/min}$$

(4) The total amount of air (dry air + moisture) required for the dryer is:

$$m_a^i = 684.19 + (684.19)(0.013) = 693.08 \text{ kg air/min}$$

The total amount of air (dry air + moisture) exited from the dryer is:

$$m_a^i = 684.19 + 14.737 = 698.92 \text{ kg air/min}$$

(5) The absolute humidity of the air leaving the dryer will be:

$$H = \frac{684.49 * 0.013 + 14.737}{684.19} = 0.0345 \text{ kg water/kg dry air}$$

which is slightly higher than the estimated value in Step (3) where it was estimated as 0.032 kg water/kg dry air.

(6) Using the psychrometric chart, a specific volume of 1.075 m^3/kg dry air is obtained for the inlet air, so the volumetric air flow rate is:

$$(684.19)(1.075) = 735.50 \ m^3/\text{min}$$

(7) The drying mechanism is similar to the tray drying in Chapter 7, which includes constant and falling rates. The estimation of total drying time can be obtained from Equation (8.11):

$\rho_L = 1100 \text{ kg/m}^3$
$d_0 = 65 \ \mu\text{m} = 6.5 \times 10^{-5} \text{ m}$ (using the largest droplet size for the longest drying time)
$\lambda = 2360 \text{ kJ/kg}$
$k_g = 0.0307 \text{ w/m K}$ (read from Table A3 of Heldman and Singh [3])

$T_a = 100°C = 373\ K$
$T_w = 36.5°C = 309.5\ K$ (from the psychometric chart)
$\rho_p = 1450\ kg/m^3$ (assumed)
$w_c = 48\%$ (wet basis) $= 0.923$ kg water/kg dry solid
$w_e = 5\%$ (wet basis) $= 0.0526$ kg water/kg dry solid
$\Delta T_{ave} = (373 - 309.5)/2 = 31.75\ K$
d_c Unknown (needs further calculation)

(8) The droplet diameter at the end of the constant rate period can be computed based on the conservation of dry solids in the drying process. The amount of dry solid in a droplet (60 µm) is: (volume) (density) (solid content). At the beginning of the constant rate drying period:

$$(4/3)\pi\,(32.5 \times 10^{-6})^3\,(1100)\,(0.25) = 3.945 \times 10^{-11}\ kg\ dry\ solid$$

At the end of the constant rate drying period:

$$(4/3)\pi\,(d_c/2)^3\,(1450)\,(0.52) = 394.793\,d_c^3\ kg\ dry\ solid$$

Therefore

$$3.954 \times 10^{-11} = 394.793 d_c^3$$

so

$$d_c = 4.64 \times 10^{-5}\ m = 46.4\ µm$$

(9) Substituting all parameters into Equation (8.11):

$$t = \frac{\rho_L \lambda d_0^2}{8 k_g (T_a - T_w)} + \frac{\rho_p d_c \lambda (w_c - w_e)}{6 h \Delta T_{ave}}$$

$$= \frac{1100 * 2360 * 1000 * (6.5 \times 10^{-5})^2}{8 * 0.0307(373 - 309.5)} +$$

$$\frac{1450 * 4.64 \times 10^{-5} * 2360 * 1000 * (0.923 - 0.0526)}{6 * (2 * 0.0307 / 4.64 \times 10^{-5}) * 31.75}$$

$$= 0.715 + 0.548$$

$$= 1.26\ sec$$

(10) Because the height and diameter of the chamber are dependent on each other, a height to diameter ratio of 3:2 is assumed, so:

$$L/D = L/(2r) = 3/2$$

or

$$r = \frac{1}{3}L$$

(11) The mean air speed can be calculated via the air flow rate of 735.50 m³/min:

$$\bar{u} = \frac{735.5}{\pi r^2} = \frac{735.5}{\pi\left(\frac{1}{3}L\right)^2} = (2107.05)/L^2[\text{m}/\text{min}] = (35.11)/L^2[\text{m}/\text{sec}]$$

(12) To ensure sufficient residence time, the height of the chamber is:

$$L = (35.11)/L^2 * (1.26)$$

so

$$L = 3.54 \text{ m and the diameter of the chamber is } 2.36 \text{ m}$$

(13) Chamber dimensions of 2.36 m diameter and 3.54 length would be initial values of the design. In Step 10, instead of restricting the length to diameter ratio, assume a chamber diameter of 2 m. In this case, the average air velocity in the chamber is:

$$\bar{u} = \frac{735.5}{\pi r^2} = \frac{735.5}{\pi(1)^2} = 244.11 \text{ m/min} = 3.9 \text{ m/sec}$$

and the height of the chamber would be:

$$L = 3.9 * 1.07 = 4.17 \text{ m}$$

In either case these height and diameter chamber dimensions can only be initial values of design because factors such as airflow patterns, ambient temperature, and feed rate would cause variations that must be considered in a more detailed analysis.

8.4 EXPECTED RESULTS

(1) Run the spray dryer at three different air inlet temperatures, three speeds in the centrifugal atomizer, and three levels of air pressure in the two-fluid nozzle atomizer. Select three of these experiments and perform a heat and mass balance.

(2) Compute the overall and evaporative efficiency as well as the residence time for all the experiments.

(3) Obtain the particle size distribution of the final product for each experiment with the centrifugal atomizer operated at different speeds.

8.5 QUESTIONS

(1) What differences did you observe (if any) in the solubility of the powder when varying the feed solids content? Discuss.

(2) What differences did you find (if any) between using the centrifugal and nozzle atomizer?

(3) How does the speed of the atomizer (or air pressure) affect the particle size distribution of the powder? How does it affect product solubility?

(4) How does the TS affect the solubility of the final product?

(5) Discuss advantages and disadvantages of different ways to increase the capacity of your spray dryer (increasing inlet temperature against decreasing outlet temperature).

8.6 REFERENCES

1. Masters, K. 1972. *Spray Drying: An Introduction to Principles, Operational Practice and Application.* Leonard Hill; London.
2. Hall, C. W. and T. I. Hedrick. 1971. *Drying of Milk and Milk Products.* AVI, Westport, CT.
3. Heldman, D. R. and R. P. Singh. 1980. *Food Process Engineering.* 2nd ed. AVI, Westport, CT.
4. Siebel, J. E. 1982. Specific heat of various products. *Ice Refrig.* 2:256.

Drying of Foods: Part III. Freeze Drying

9.1 INTRODUCTION

Freeze drying is a unit operation consisting of the removal of moisture from a product by sublimation of free water from the solid phase. Normally this process results in the highest overall quality for a dehydrated product as a result of extremely low temperatures during drying. The low temperatures associated with freeze drying result in less occurrence during drying of various undesirable thermally activated reactions, and the rigid structure of the frozen product mechanically prevents shrinkage during the drying process, allowing for excellent rehydration properties. Characteristics of final products are good flavor, aroma, nutrient retention, and excellent rehydration properties. However, the process is very slow and expensive compared with other types of drying because of its energy consumption and use of expensive equipment over long periods of time.

During freeze drying, ice is sublimed to vapor that is removed from the system. To accomplish this, pressure is reduced below the triple point (see Figure 9.1), and temperature is increased at a constant low pressure. This will cause a change of phase from solid directly to vapor without passing through the liquid phase. As the product is dried, a sharp break between the dried and frozen portions is formed. This line, known as the sublimation front, is assumed to recede uniformly within the product. The channels left by the subliming ice provide the paths for vapor to escape from the active drying zone [2,3].

The basic setup of a simple vacuum freeze-drying system is shown in Figure 9.2. The typical freeze-dryer includes an airtight chamber, vacuum pump to obtain high vacuum in the chamber, vapor trap, and a condenser cooled by a refrigerant to desublime water back to ice. The frozen product is placed on trays that can be heated by the circulation of a heating fluid to provide energy for vaporization.

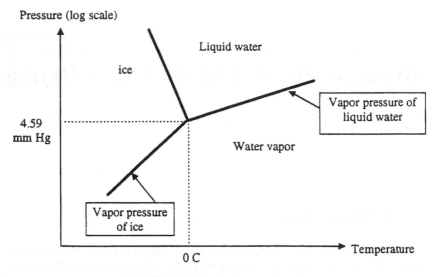

FIGURE 9.1. Phase diagram for water.

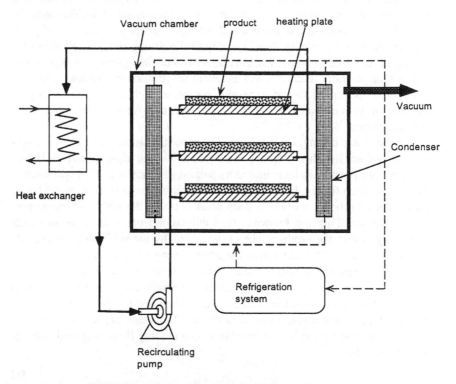

FIGURE 9.2. Schematic diagram of a freeze-dryer system.

At the beginning of the freeze-drying process, drying rates are high because there is little resistance either to heat transport from the plate to the material or to mass flux from the material to the condenser. However, as drying proceeds, a porous, dried, and highly resistant layer builds up around the material. The dried layer offers more resistance to the heat flux than the frozen layer, but at low pressure is less resistant to mass transfer. Because heat and mass transfer resistance are functions of pressure, increasing the pressure may enhance heat transfer at the expense of an increased resistance to mass transfer. It is thus clear that the chamber pressure is a major control variable of the freeze-drying process. The temperature of the plates is another important variable because it affects the rate of heat transmission to the surfaces of the material being dried and the energy reaching the interface between the dried and frozen layers. The condenser temperature is also a control variable because it affects the driving force of the water vapor for mass transfer.

The mode of heat transfer depends on the arrangement of samples in the chamber. The simplest case is only radiation to both upper and lower surfaces. However, there can also be radiation to the upper dried surface and conduction through a film layer at the lower surface. This laboratory exercise will give students practical experience with the freeze-drying process.

9.2 OBJECTIVES

(1) To become acquainted with the freeze-drying phenomenon
(2) To determine how sample thickness affects the drying process
(3) To observe how different modes of heat transfer affect the rate of drying
(4) To observe the quality of freeze-dried products
(5) To determine how some of the main control variables affect the rate of heat and mass transfer to and from drying materials

9.3 MATERIALS AND METHODS

9.3.1 Materials

The following is a list of materials required for the experiments described in this chapter:

- a freeze-dryer consisting of a drying chamber, vacuum pump, and condenser (the equipment should have appropriate accessories to control the temperature of the heating plates and record the temperature of the samples)

- a blast freezer for quick freezing of samples prior to drying
- a complete setup for moisture determination consisting of an electronic balance with a 1-mg resolution and a vacuum oven
- three different raw materials to be dried (suggested: apples, beef, and carrots)
- a PC or equivalent for logging data (a board with a minimum of six channels is required)
- appropriate trays to place samples in the chamber (one of the trays should allow the product to receive heat by radiation on both surfaces)

9.3.2 Methods

(1) Prepare the samples to be frozen: pretreatment such as a heat or chemical treatment may be done prior to drying to prevent enzymatic reactions from occurring during the drying process. Cut the samples into slices of several different thicknesses, measuring and recording the varying sizes. Prepare enough raw material to load the drying chamber three times with different temperature levels set for each heating tray.

(2) Freeze the samples in a blast freezer at –20°C until completely frozen (approximately 1.5 hr), and then set some of the samples aside for moisture determination after the experiment is completed.

(3) While the samples are in the blast freezer, prepare the freeze-dryer for operation by gathering a set of six thermocouples and connecting them to the front panel. (Prior calibration may be conducted by placing them in ice from distilled water followed by boiling water. Be sure to note the total pressure over the boiling water.)

(4) Once the samples are frozen, place them on the trays and then into the drying chamber. Select two types or arrangements—one to receive radiation on both surfaces and the other to receive heat by conduction on the bottom and radiation on the top. The latter case will be run as a reference to compare against the radiation only arrangement.

(5) Place two thermocouples on each product to measure the surface temperature of the samples. Three different food products should be loaded simultaneously for each run of the dryer.

(6) Following the operating instructions for the equipment, close the chamber, pull the air out of the chamber, turn the refrigerating system on, select an appropriate temperature (between 20 and 40°C) for the heating plates, and turn them on. (Be aware of the collapse temperature of the material being dried.)

(7) During drying, record the surface temperature of the samples, condenser

temperature, drying chamber pressure, and heating plate temperatures at 10-min intervals.

(8) Freeze drying normally requires long processing times, so once the experiment is running, the computer automatically records the sample temperatures. It is preferable to run the experiment as long as the computed theoretical time and occasionally monitor the experiment. Collect the dried samples when the experiment is complete.

(9) Finally, determine the moisture content of the raw materials and dehydrated product by weighing the samples, placing them in a vacuum oven at 75°C, and turning the pump on. Allow the samples to remain at least 24 hr, making sure the absolute pressure does not exceed 50 mm Hg. Weigh the samples after drying, and determine the moisture content as a percentage of initial weight.

9.3.3 Calculations of Freeze-Drying Times

Several approaches have been proposed for the computation of drying time for freeze dehydration; the URIF model (uniformly retreating ice front) has been quite successful for this type of prediction and analysis. Figure 9.3 shows the URIF model as applied to slab geometry with drying from both sides. It is assumed that the partial pressure in the refrigeration evaporator (p_c) is equal to that at the surface of the product (p_o) assuming no leaks in the chamber. In terms of mass transfer, the mass flux flow rate per unit area of the water vapor G_s is given by [1]:

$$G_s = \frac{K_p\left(p_s p_o\right)}{z} = \frac{K_p\left(p_s - p_c\right)}{z} \tag{9.1}$$

$$G_s = \rho\left(\frac{x_0 - x_f}{1 + x_0}\right)\frac{dz}{dt} \tag{9.2}$$

where

p_s = vapor pressure of water at the sublimation front
K_p = constant related to the permeability of the dry layer
ρ = density of the frozen material
x_o = initial moisture content
x_f = final moisture content
z = thickness of the dry layer
t = drying time

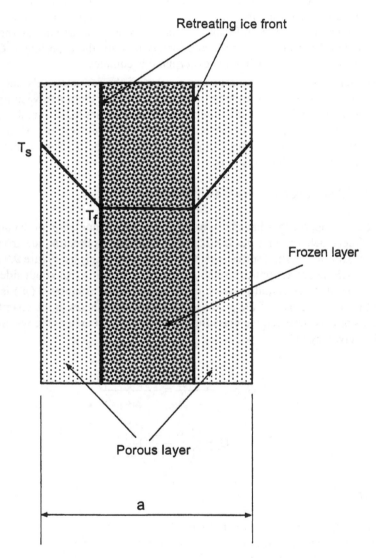

FIGURE 9.3. URIF model applied to a slab geometry.

Assuming that p_s, p_c, and K_p are constants, equate Equations (9.1) and (9.2) and integrate to obtain:

$$t = \frac{\rho\left(x_o - x_f\right)\overline{H}_s a^2}{2\left(1 + x_o\right)k_T\left(T_0 - T_s\right)} \qquad (9.3)$$

where the additional symbols are defined as:

k_T = thermal conductivity
$\overline{H}_s$ = average latent heat of evaporation

To calculate the conditions at the sublimation front, two equations are used, the first resulting from combining the heat and mass transfer relationships demonstrated in Equation (9.3):

$$p_s - p_o = -\frac{k_T}{K_p \overline{H}_s}\left(T_s - T_o\right) \qquad (9.4)$$

The second equation is the Clausius-Clapeyron relationship:

$$\ln\left(p\right) = 30.9526 - \frac{6153}{T} \qquad (9.5)$$

where p is the absolute pressure in μHg and T is the temperature in K.

The vapor pressure and temperature at the sublimation front are the two unknowns in this system, but a solution can be found graphically (as revealed in Example 2).

9.3.4 Example 1

Estimate the freeze drying time for a 25-mm-thick slab of beef whose initial density is 1070 kg/m³, initial moisture content 75% (wet basis), and dry state moisture content 4% (wet basis). Maintain the sublimation pressure at 250 μHg and the condenser pressure at 100 μHg, with the K_p value at 2.3 × 10^{-9} kg/m μHg. What would the freeze-drying time be if the slab has a thickness of 15 mm?

9.3.4.1 SOLUTION

(1) $a = 0.0125$ m
   $x_o = 0.75/0.25 = 3.0$ kg water/kg dry solid
   $x_f = 0.04/0.96 = 0.0417$ kg water/kg dry solid

$$P_o = P_c = 100 \ \mu Hg$$
$$P_s = 250 \ \mu Hg$$

Substitute these parameters into Equation (9.3) to obtain:

$$t = \frac{\rho\left(x_o - x_f\right)a^2}{2K_p\left(1 + x_o\right)\left(p_s - p_o\right)}$$

$$= \frac{(1070)(3.0 - 0.0417)(0.0125)^2}{(2)(2.3 \times 10^{-9})(1 + 3.0)(250 - 100)}$$

$$= 179,200 \ sec$$

$$= 49.8 \ hr$$

(2) All parameters are the same as in Part (1) except $a = 0.0075$ m.

$$t = \frac{\rho\left(x_o - x_f\right)a^2}{2K_p\left(1 + x_o\right)\left(p_s - p_o\right)}$$

$$= \frac{(1070)(3.0 - 0.0417)(0.0075)^2}{(2)(2.3 \times 10^{-9})(1 + 3.0)(250 - 100)}$$

$$= 64,511 \ sec$$

$$= 17.9 \ hr$$

It is evident that the thickness has a great influence on drying time.

9.3.5 Example 2

Take a slab of beef 15 mm thick, 1070 kg/m³ dense, with an initial moisture (wet basis) of 70% and a dried moisture of 3% and set the temperature at the surface of the slab at 40°C and the condenser at –35°C. Assuming the pressure in the condenser is equal to the pressure at the surface (no losses), estimate the drying time.

9.3.5.1 SOLUTION

$K_p = 0.7 * 10^{-9}$ kg/m sec μHg (permeability of the dry layer)
$k_T = 0.035$ W/m K (thermal conductivity of the dry layer)

To determine p_c, use Equation (9.4) for $T_c = -35°C$

$$\ln\left(p_c\right) = 30.9526 - \frac{6153}{(-35 + 273.3)}$$

then $p_c = 285\ \mu Hg = p_o$ (assumed).

One point on the combined heat and mass transfer line (see Figure 9.2) is now known as $(T_o P_o) = (-35°C, 285\ \mu Hg)$.

The slope of the combined heat and mass transfer line representing the process is:

$$-\frac{k_T}{K_p \overline{H}_s} = \frac{0.035}{(0.7 * 10^{-9})(2.849 * 10^{-6})} = -17.55 \frac{\mu Hg}{K}$$

Using the slope of the line and the known point, one can graph the line in a pressure versus temperature coordinate system. On the same graph, plot the Clausius-Clapeyron [Equation (9.5)], as shown in Figure 9.4.

The intersection of the two operating lines and equilibrium line will be the values T_s and p_s.

$p_s = 1180\ \mu Hg$
$T_s = -15.8°C$

Plugging in these values and knowing that

$x_o = 70$ kg water/30 kg dry solids = 2.3333 kg water/kg dry solids
$x_f = 3/97 = 0.030927$ kg water/kg dry solids
$a = 0.015/2 = 0.0075$ m

one may substitute Equation (9.3) to obtain:

$$t = \frac{1073 * (2.333333 - 0.030927) * 0.0075^2}{2 * (0.7 * 10^{-9})(1178 - 285)}$$

$$= 1.1115 \times 10^5\ sec$$

$$= 31\ hr$$

It is important to emphasize that the values k_T and K_p are characteristic of each material. Notice also that the higher the temperature the faster the process can go, but there is a limiting temperature known as the collapse temperature for each material. It is always desirable to work below that value to obtain a quality product.

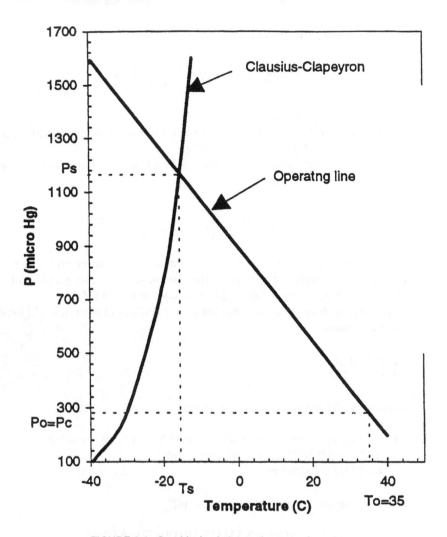

FIGURE 9.4. Graphical solution to the example problem.

9.4 EXPECTED RESULTS

(1) The student is expected to compute the theoretical time required to reach the final moisture content obtained in each experiment and compare it against the experimental value.

(2) Show the drying curve and drying rate curve for all the experiments (see Chapter 7 for a detailed description of these plots).

9.5 QUESTIONS

(1) Did the computed theoretical time agree with the experimental value obtained in the lab? If not, discuss the possible reasons for the disagreement. What sources of error may occur during the experiments? Which of the assumptions in the theoretical analysis do you consider invalid?

(2) Because freeze drying is an energy intensive process, which parameters do you think can be changed in a commercial operation to improve the heat and mass transfer during drying? Discuss the main operating variables (heating plate and condenser temperature and vacuum of the drying chamber). How does each variable affect the overall thermal efficiency of the freeze-dryer?

(3) Observe and taste the dehydrated product. Comment on the properties (i.e., flavor, texture, aroma) of the materials you dried, and compare them with other dehydrated products you obtained with other drying techniques.

9.6 REFERENCES

1. Goldblith, S. A., Rey, L., and Rothmayr, W. W. 1975. *Freeze Drying and Advanced Food Technology.* Academic Press; New York.
2. Toledo, R. T. 1991. *Fundamentals of Food Process Engineering.* 2nd ed. Van Nostrand Reinhold, New York.
3. King, C. J. 1970. Freeze-drying of foodstuffs. *Critical Reviews Food Technology.* 1:379.
4. King, C. J. 1973. Freeze Drying. In *Food Dehydration,* 2nd ed., vol. 1. W.B. Van Arsdel, M. J. Copley, and A. I. Morgen (Editors). AVI Publishing Co., Westport, CT.

CHAPTER 10

Extrusion of Foods

10.1 INTRODUCTION

Extruded food products have gained increasing importance during the last few decades. Ready-to-eat (RTE) cereals and pasta products have been the main application of extrusion during early stages of this technology, but now it has expanded into the preparation of dry and semimoist pet foods, snacks, confectionery products, chewing gums, modified starches, dry soups and beverages, and texturized vegetable proteins.

No single definition can be given to extrusion. It can be considered a shaping process by forcing a product through a die, a high-temperature-short-time (HTST) thermal process, a cooking technique, or a bioreactor in which chemical reactions involving natural biopolymers of a food system take place. Some of the advantages of extrusion described by Harper [2] are its versatility, high productivity, low cost, ability to form different shapes, high microbiological quality, and nutrient retention due to the HTST process. It is also energy efficient and causes almost no environmental impact.

Several types of extruders have been developed for food processing; the simplest ones contain a single screw, and the increasingly complex ones possess various twin screw arrangements. In terms of functional characteristics, the different extruders can be classified as those for pasta forming, high pressure, low shear cooking, and high shear cooking. This laboratory practice will focus on the study of the performance of a single screw device. The extrusion equipment [2, 3] is divided into feeding, transition, and metering sections. (Figure 10.1 shows a schematic for a single screw that identifies these three parts.)

In the feeding section of the extruder, the material is fed from a hopper. Several designs have been developed to ensure the product is feeding continuously

105

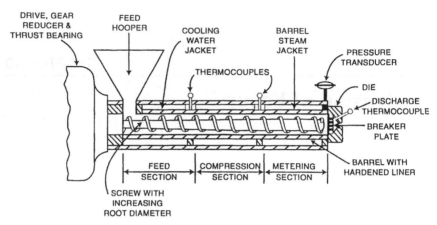

FIGURE 10.1. Schematic of an extruder (reprinted from Reference [6], p. 67, with permission of IFT).

through a screw-barrel arrangement. The equipment for this laboratory practice might have a self-driven screw that forces the product through the extruder so screw length remains uniform; other screw types have a different pitch in their feeding section to obtain consistent product output.

In the transition or compression section the raw material is plasticized as the mixture is heated by mechanical friction and heat transfer from the walls of the barrel. Dramatic physical and chemical changes and a mixing effect take place in this section when temperature and pressure increase as the product progresses down the barrel. The chemical reactions are affected by the exclusion of most of the oxygen during the extrusion process [1], and the physical changes are a consequence of the high shear rates produced with the screw rotation.

The barrel is designed to avoid burning and slipping of the product at the wall which would build pressure in the machine and drag flow. Other important parameters affecting the screw's performance are ratio of the length to diameter (L/D), compression ratio, pitch height, shape of the flight, and clearance between the screw and barrel.

The metering section is located just before the extruder's discharge. As shown in Figure 10.1, the highest pressure is obtained in this section. Almost every extruder works as an inefficient pump because most of the energy supplied is used in building pressure against a die located at the product exit. The die also plays a role in shaping the final product and provides a sudden expansion that has a flashing effect on the food as it evaporates some of the water trapped in the food matrix. The product properties (elasticity, hardness, hydration ability, density, morphology, and degree of texturization) may be affected by both process parameters (pressure and temperature profiles, thermal and shear loads, residence time distribution, moisture correction, mixing, and chamber or channel filling) and feed supply properties (choice of material, moisture and fat content, purity, particle size distribution, and rheology).

10.2 OBJECTIVES

(1) To achieve familiarity with a typical extrusion operation
(2) To study the effect of shear rate, product pressure, and temperature on the characteristics of an extruded food product
(3) To observe the influence of different die designs on the operational characteristics and final product
(4) To determine the influence of product moisture on the operation of an extruder and in the final product

10.3 MATERIALS AND METHODS

10.3.1 Apparatus and Materials

- a 3/4" C. W. Brabender laboratory extruder with L/D = 20:1 or equivalent and a compression ratio of 2:1 (The extruder is driven by a Brabender DO-CORDER 'E' torque rheometer whose objective is to measure the torque applied to the screw when it pumps the product. This measure gives an indication of the rheological properties of the dough. The Brabender DO-CORDER 'E' is also provided with appropriate temperature instruments to control and monitor the extrusion operation.)
- a feed hopper moved by a 1/20 HP electrical motor
- six dies of different diameters (1/32, 1/16, 3/32, 1/8, 5/32, and 3/16")
- a pressure transducer to measure the product pressure in the metering section
- thermocouples for temperature measurement
- a specific preparation recipe (A suggested product is described below, although the student is encouraged to try alternatives given the extrusion has numerous applications for several types of food.)
  — 5 kg wheat flour
  — 12 eggs
  — 3 kg sugar
  — 0.3 kg salt
  — 0.1 kg baking soda
  — water

10.3.2 Procedure

(1) *Assembly:* Before startup, assemble the extruder and related equipment. Some food extruders are designed so that the screw and barrel are in segments that must be assembled before the correct operating configuration can be achieved. Normally each extruded product has its own configuration, and the marked parts are laid out systematically to achieve proper assembly. Prior to startup, all feeders and regulators need to be checked to determine if they are operating properly.

(2) *Startup:* The purpose of the startup is to bring the extruder to operating condition and equilibrium as soon as possible. Prior to feeding ingredients into the extruder, bring the system as close as possible to a steady-state operating temperature. For a cooking extruder, this usually means the introduction of steam into the jackets surrounding the barrel or to the barrel itself and into the preconditioning chamber if one is being used.

(3) *Dough Preparation:* Prepare the dough following the recipe listed above or one of your own, varying the proportions of water added (10, 30, and 50% by weight) in three different batches.

(4) *Feeding:* Following the operating instructions for the extruder and torque rheometer, feed the product into the hopper. Wait until you observe a steady-state operation and prepare a data sheet with the following:
- identification of the sample
- rpm at which the screw is turning
- torque applied to the screw when running full of product (Adjust the rpm of the feed hopper until there is a constant torque, because fluctuations usually mean the feed to the extruder is not uniform.)
- pressure of the product at the metering section using the pressure transducer
- temperature of the product as it passes through the die
- product swell, obtained as the ratio of the die orifice diameter over the diameter of the extruder
- rate of extrusion expressed as mass/min or length/min
- any other important observations related to the extruded product (such as visual appearance and texture)
- moisture of the final product obtained in the vacuum oven (see Chapters 7 to 9)

(5) *Running the Experiment:* Run the experiment several times for each product, selecting at least three processing temperatures (140, 180, and 250°C) in the barrel, three levels of shear rate (50, 100, and 250 rpm), and two die diameters (1/32 and 3/16″), noting that the values given here are only suggestions.

10.3.3 Calculations

The engineering aspects of the extrusion process are quite complex because most of the products are non-Newtonian. Therefore, a detailed analysis will not be presented here. For the student interested in more information, reading References [2], [3], and [4] is highly recommended.

Pumping efficiency is defined [4] as the fraction of motor energy required to move a material through an extruder and generate an appropriate pressure. It can be expressed as:

$$\eta = \frac{E_p}{E} \tag{10.1}$$

The energy used in pumping E_p can be obtained from:

$$E_p = Q_v * P \tag{10.2}$$

where Q_v is the volumetric flow rate and P the pressure at the die. The energy applied by the electric motor to the extruder can be easily obtained from the rotating speed (ω) and torque (T):

$$E = \omega * T \tag{10.3}$$

10.4 EXPECTED RESULTS

(1) Submit a data sheet with all the variables described in Section 10.3.2 that includes the pumping efficiency for each experiment and any significant observations on product quality.

10.5 QUESTIONS

(1) How does the extruder rotating speed affect the pumping efficiency of the process?
(2) Is there any effect from the moisture content of the product on the torque applied to the screw? Why? Does it affect the pumping efficiency?
(3) How is the final product affected by the maximum pressure in the extruder? How is it affected by the temperature in the transition section?
(4) How does the die size affect the operational characteristics of the extruder (torque, pressure) and the final product?

10.6 REFERENCES

1. Hoseney, C. R. 1984. Chemical changes in carbohydrates produced by thermal processing. *Journal of Chemical Education.* 61:308–312.
2. Harper, J. M. 1981. *Extrusion of Foods. Volume 1.* CRC Press, Boca Raton, FL.
3. Harper, J. M. 1989. Food extruders and their applications. In *Extrusion Cooking.* C. Mercier, P. Linko, and J. M. Harper, eds. American Association of Cereal Chemists, St. Paul, MN.
4. Janssen, J. L. 1989. Engineering aspects of food extrusion. In *Extrusion Cooking.* C. Mercier, P. Linko, and J. M. Harper, eds. American Association of Cereal Chemists, St. Paul, MN.
5. Levine, L. 1992. Extrusion Processes. In *Handbook of Food Engineering.* D. Heldman and P. Maraylatara, eds. McGraw-Hill, New York.
6. Harper, J. M. 1978. Extrusion processing of food. *Food Technol.* 32(7):67–72.

CHAPTER 11

Evaporation

11.1 INTRODUCTION

Evaporation is a concentration process of a solution that boils off solvent. One of its primary objectives is bulk and weight reduction of fluids, which permits more efficient transportation of important product components and storage of the solid. Another important objective of evaporation is to remove large amounts of moisture effectively and efficiently before a food material enters a dehydration process (e.g., milk powder manufacture). The evaporation process is also used to reduce water activity by increasing the concentration of soluble solids in food materials as an aid to preservation (e.g., in sweetened condensed milk manufacture). The evaporation process has been extensively used in the dairy industry to concentrate milk; in the fruit juice industry to produce fruit juice concentrates; and in the manufacture of jams, jellies, and preserves to raise sugar solutions for crystallization. Evaporation can also be used to raise the solids of dilute solutions prior to spray or freeze drying.

The process of evaporation involves the application of heat to vaporize water at the boiling point. Because of the temperature sensitivity of most food products, a relatively high vacuum (low pressure) is used to make water boil and evaporate at a lower temperature. The basic factors that affect the rate of evaporation are as follows: (1) the rate at which energy can be transferred to the liquid, (2) the quantity of energy required to evaporate each kilogram of water, (3) the maximum allowable temperature of the liquid, (4) the pressure at which the evaporation takes place, and (5) any changes that may occur in the foodstuff during the course of the evaporation process.

An evaporator system normally consists of four basic components [1] (as illustrated in Figure 11.1):

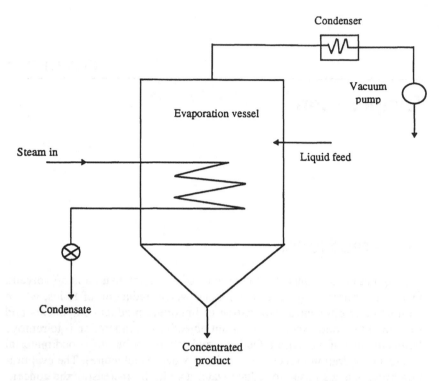

FIGURE 11.1. Schematic diagram of evaporation (reprinted from Reference [4], p. 438, with permission of Chapman & Hall).

(1) An evaporation vessel functioning as a reservoir for the product and allowing separation of vapor

(2) A heat source to supply sensible and latent heat of evaporation to the feed

(3) A condenser to condense the vapor

(4) A method of maintaining vacuum (removing non-condensible gases)

An evaporator is essentially a heat exchanger in which a liquid is boiled to give a vapor, making it a low-pressure steam generator as well. It may be possible to treat an evaporator as a low-pressure boiler while making use of the steam produced for further heating in another evaporator. Two evaporators connected in a series is called two-effect evaporation, three evaporators connected in a series is called three-effect evaporation, and so forth. It is clear that multiple-effect evaporators are energy efficient because energy is used repeatedly. However, this energy efficiency is at the cost of increased capital for the evaporator. The n effects will cost approximately n times as much as a single effect [2].

Comparative costs of auxiliary equipment do not follow the same pattern because there are fewer condenser requirements for multiple-effect evaporators. The optimum design for an evaporation plant must therefore be based on a balance between operating costs (which are lower for multiple effects because of their reduced steam consumption) and capital charges (which will be lower when fewer evaporator effects are required).

11.2 THERMODYNAMICS OF EVAPORATION

Thermodynamics is important in the description of the evaporation process for food products because a phase change is involved when water in food material becomes vapor. As evaporation proceeds, boiling temperatures rise because the remaining liquids become more concentrated. The rise in temperature reduces the available temperature drop if it is assumed that the heat source undergoes no change, and the total rate of heat transfer will drop accordingly.

11.2.1 Phase Change

During the evaporation process, the phase change is evaporation of water from the liquid to the vapor state. The latent heat of vaporization of a liquid food can be described as a function of pressure by a modified Clausius-Clapeyron equation [1]:

$$\ln\left(p'\right) = \frac{L_v'}{L_v} \ln\left(p\right) + C \ln\left(p\right) + C \tag{11.1}$$

where

> p = vapor pressure of pure water
> L_v = latent heat of vaporization of pure water
> p' = vapor pressure of liquid foods
> L_v' = latent heat of vaporization of water in liquid foods
> C = constant

By plotting the natural logarithm of vapor pressure for a liquid food product versus the natural logarithm of the vapor pressure for pure water at various temperatures, the relationship between the latent heat of the liquid food product and the latent heat of water is established. Information of this type is necessary to account for the change in latent heat as the concentration of a fluid food increases during evaporation.

11.2.2 Boiling Point Elevation

As evaporation proceeds, the liquor remaining in the evaporator becomes more concentrated and its boiling point rises. The extent of the boiling point elevation depends on the nature of the material being evaporated and the concentration changes produced. The extent of the rise can be predicted by the following expression [1]:

$$\frac{\lambda_v}{R_g}\left(\frac{1}{T_{A0}} - \frac{1}{T_A}\right) = -\ln\left(X_A\right) \tag{11.2}$$

where

λ_v = latent heat of vaporization
T_{A0} = boiling point of pure water
T_A = boiling point of the solution
X_A = mole fraction of water in the solution
R_g = universal gas constant

Assuming that the boiling point elevation is small and only the first term of the logarithmic expansion of Equation (11.2) is used, the following expression is obtained:

$$\Delta R_B = \frac{R_g T_{A0}^2}{\lambda_v} X_B \tag{11.3}$$

where

X_B = mole fraction of solute causing the boiling point elevation
ΔT_B = elevation of the boiling temperature

Introducing the concept of molality (m) into Equation (11.3), the following expression results:

$$\Delta T_B = \frac{R_g T_{A0}^2 W_A}{1000\lambda_v} m \tag{11.4}$$

where

λ_v = latent heat of vaporization per unit mass of water
W_A = molecular weight of water
m = molality of the solute (moles of solute/1000 g H_2O)

Equation (11.4) can be utilized to compute the boiling point elevation as long as the solution is dilute. In a situation in which the product may become highly concentrated (as will occur in the evaporation process), the assumption made in obtaining Equation (11.4) may create considerable error. In a higher concentration, Equation (11.2) should be used to compute the boiling point of the product.

The second method commonly used is to estimate boiling point elevations based on Duhring's rule which states that the temperature at which one liquid exerts at a given vapor pressure is a linear function of the temperature at which a reference liquid exerts on identical vapor pressure. Thus, if we take the vapor pressure (temperature relation of a reference liquid—usually water) and know the vapor pressure at two temperatures (temperature curve of the solution that is being evaporated), the boiling points of the solution to be evaporated at various pressures can be read from the diagram known as the Duhring plot that gives the boiling point of solutions at various concentrations by interpolation at various pressures along a line of constant composition. (A Duhring plot of boiling points for sodium chloride solutions is given in Figure 11.2.)

11.2.3 Heat and Mass Balance During Evaporation

A mass balance conducted on the evaporation chamber shown in Figure 11.1 results in the following overall mass balance:

$$F = V + P \tag{11.5}$$

where

F = feed (kg/sec)
V = vapor (kg/sec)
P = product (kg/sec)

The material balance on the liquid is:

$$X_F F = X_P P \tag{11.6}$$

where

X_F = solid fraction in the feed
X_P = solid fraction in the product

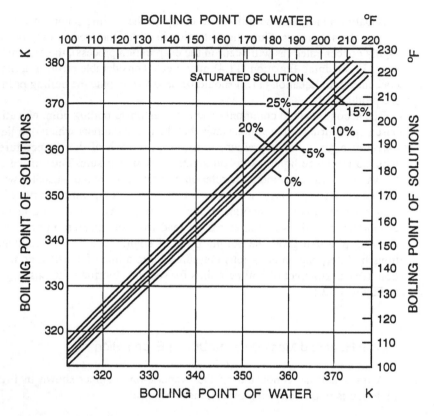

FIGURE 11.2. The influence of solute concentrations of boiling point elevation of NaCl (reprinted from Reference [1], p. 222, with permission of Chapman & Hall).

An enthalpy balance conducted on the evaporation chamber shown in Figure 11.1 results in the following:

$$Fc_{pF}(T_F - 0) + w_sH_s = VH_v + Pc_{pP}(T_P - 0) + w_sH_c \qquad (11.7)$$

where

c_{pF} = specific heat of the feed
c_{pP} = specific heat of the product
T_F = temperature of the feed
T_P = temperature of the product in the evaporator
w_s = steam supply
H_s = enthalpy of steam
H_c = enthalpy of condensate
H_v = enthalpy of vapor

Here a base temperature of 0°C is used to obtain the enthalpy values from standard steam tables. The heat added by cooling the condensate below the condensing temperature will be small and is neglected in most computations, as is heat loss from surfaces of the evaporator body. The heat transfer from the heating medium to the product is described by the following equation [1]:

$$q = UA\left(T_s - T_p\right) \tag{11.8}$$

where

U = overall transfer coefficient
T_s = temperature of steam
T_p = temperature of the product

The heat needed to be transferred from steam to the product for accomplishing the specified extent of evaporation is given by the following equation:

$$q = w_s H_s - w_s H_c \tag{11.9}$$

The efficiency of the evaporation process is the steam economy given by the following equation:

$$\text{steam economy} = V/w_s \tag{11.10}$$

which expresses the mass of water evaporated from the product per unit mass of steam utilized. The objective of this laboratory practice is to provide hands-on experience with a single-effect evaporator.

11.3 OBJECTIVES

(1) To achieve familiarity with a typical evaporation process
(2) To study the effect of external pressure and dissolved solute on the boiling point of a solution

11.4 PROCEDURES

Pilot plant and bench top scale are the two options designed for this laboratory practice. If the first is not available, the second may be constructed.

11.4.1 Materials and Apparatus

- single-effect evaporator
- 4-liter round flask with flat bottom
- water ejector
- water (or oil) bath
- raw milk

11.4.2 Procedures

For the pilot plant scale evaporator:

(1) Measure the solid concentration of raw milk by the vacuum drying method.

(2) Set up the evaporation system as shown in Figure 11.3.

(3) Warm up the system with water.

(4) Turn off the steam and drain the water from the evaporator vessel.

(5) Feed the evaporator vessel with preweighed raw milk and turn on the steam.

(6) Record the initial temperature of the raw milk.

(7) When the operation is under steady state, record the pressure of the evaporation vessel and steam consumption rate and temperature.

(8) Monitor the temperature of the milk in the evaporation vessel and determine if there is a rising boiling temperature.

(9) When the concentration of milk is about 40% total solid by weight, stop evaporation by turning off the steam.

(10) Weigh the concentrated milk.

(11) Measure the solid concentration of the product using the vacuum drying method.

(12) Perform mass and energy balance.

For the bench top scale:

(1) Measure the solid concentration of raw milk by the vacuum drying method

(2) Construct the evaporation apparatus as shown in Figure 11.4.

(3) Heat the water bath to 95°C.

(4) Fill the round flask (evaporator vessel) with preweighed raw milk.

(5) Record the initial temperature of the raw milk.

(6) When the operation is at steady state, record the pressure of the evaporation vessel, steam consumption rate and temperature.

(7) Monitor and record the temperature of the milk in the round flask every 5 min, watching for a rise in boiling temperature.

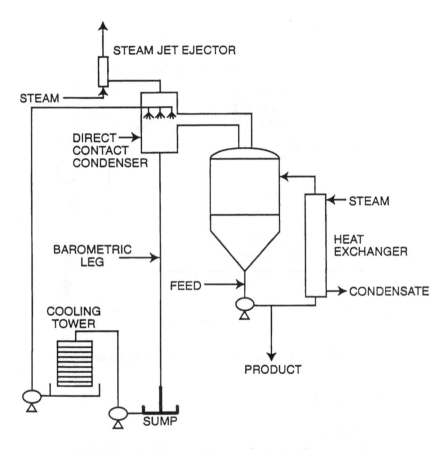

FIGURE 11.3. Schematic diagram of a single-effect evaporator.

(8) When the concentration of milk is about 40% total solid by weight, stop evaporation by turning off the vacuum and taking the round flask off the water bath.

(9) Weigh the concentrated milk.

(10) Measure the solid concentration of the product by the vacuum drying method.

(11) Perform mass and energy balance.

11.5 EXPECTED RESULTS

(1) An organized data sheet

(2) A mass and energy balance

(3) A calculated steam economy

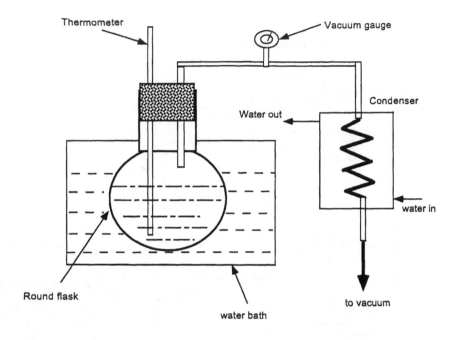

FIGURE 11.4. Schematic diagram of a bench top evaporator.

11.6 EXAMPLES

11.6.1 Example 1

Compute the boiling point rise of a 5% (w/w) NaCl solution at atmospheric pressure (sealevel).

11.6.1.1 SOLUTION

Using Equation (11.2), assume a 5% (w/w) NaCl solution.

$$\lambda_v = 2257 \text{ kJ/kg}$$

$$= 2257 \text{ kJ/kg}* 1000 \text{ J/kJ}* 18 \text{ g/mol}* 1/1000 \text{ kg/g}$$

$$= 4.0626 \times 10^4 \text{ J/mol}$$

$$m = [\text{g NaCl}/1000 \text{ g water}]/[\text{molecular weight of NaCl}]$$

$$= 50/59 = 0.8475$$

$$T_{A0} = 100.00 + 273.13 = 373.13 \text{ K}$$

$$W = 18$$

applying Equation (11.4):

$$\Delta T_B = \frac{R_g T_{A0}^2 W_A}{1000 L_v} m = \frac{(8.314 J / \text{mole} \cdot K) * (373.13K)^2 (18)}{(1000)(4.062 \times 10^4 J / \text{mol})} (0.8475)$$

$$= 0.435 \text{ K}$$

So the boiling point for the 5% NaCl would be:

$$373.13 + 0.435 = 373.565 \text{ K}$$

or

$$373.565 - 273.13 = 100.435°C$$

11.6.2 Example 2

Milk with 14% (w/w) total solids is being concentrated in a single-effect evaporator with a feed rate of 20,000 kg/hr at 10°C. The evaporator is being operated at a sufficient vacuum to allow the product moisture to evaporate at 75°C while steam is being supplied at 232.1 kPa. If the desired concentration of the final product is 40% total solids, compute the steam requirement and economy for the process when condensate is released at 75°C.

11.6.2.1 SOLUTION

(1) Perform the mass balance using Equations (11.5) and (11.6):

$$20,000 = V + P$$

$$0.14(20,000) = 0.4\, P$$

Solve for V and P:

$$P = 7000 \text{ kg/hr}$$

(2) Perform the energy balance using Equation (11.7):

$$c_{pF} = 3.852 \text{ kJ/kg K}$$

The specific heat of the product may be estimated by the following empirical equations [1]:

$$c_{pP} = c_{pS} X_{SP} + c_{pW} X_{WP} \qquad (11.11)$$

and

$$c_{pF} = c_{pS} X_{SF} + c_{pW} X_{WF} \qquad (11.12)$$

where

c_{pP} = specific heat of the product
c_{pS} = specific heat of the solid
X_{SP} = weight fraction of solid in the product
c_{pW} = specific heat of the water
X_{WP} = weight fraction of water in the product
c_{pF} = specific heat of the feed
X_{SF} = weight fraction of solid in the feed
X_{WF} = weight fraction of water in the feed

Substitute the numerical values into Equations (11.11) and (11.12):

$$c_{pP} = c_{pS}(0.4) + 4.1865(0.6)$$

$$3.852 = c_{pS}(0.14) + 4.1865(0.86)$$

Solve for c_{pP} and c_{pS}:

$$c_{pS} = 1.797 \text{ kJ/kg K}$$

$$k_{pP} = 3.23 \text{ kJ/kg K}$$

From a handbook, the following parameters are obtained:

$H_s = 2713.5$ kJ/kg
$H_v = 2635.8$ kJ/kg
$H_c = 313.93$ kJ/kg

Assume the condensate is saturated liquid and substitute the above parameters into Equation (11.7):

$$20,000(3.852)(283 - 273) + w_s(2713.5)$$

$$= 13{,}000(2635.8) + 7000(3.23)(348 - 273) + w_s(313.93)$$

Solve for w_s:

$$w_s = 14{,}665 \text{ kg/hr}$$

(3) The steam economy can be determined using Equation (11.10):

steam economy $= 13{,}000/14{,}665 = 0.886$ kg water/kg steam

11.7 QUESTIONS

(1) How will viscosity change as evaporation proceeds? What is the consequence of a viscosity change to the heat transfer coefficient during evaporation?
(2) Assuming no change in the heat source, how does a rise in boiling temperature affect the evaporation process in terms of the total rate of heat transfer?
(3) Why is the evaporation process often operated under vacuum pressure?

11.8 REFERENCES

1. Heldman, D. R. and R. P. Singh. 1981. *Food Process Engineering.* 2 ed. AVI, Westport, CT.
2. Earle, R. L. 1983. *Unit Operations in Food Processing.* 2nd ed. Pergamon Press, New York.
3. Brennan, J. G., J. R. Butters, N. D. Cowell, and A. E. Lilley. 1990. *Food Engineering Operations.* 3rd ed. Elsevier Applied Science, New York.
4. Toledo, R. T. 1991. *Fundamentals of Food Processing Engineering,* 2nd ed. Van Nostrand Reinhold, New York.

CHAPTER 12

Physical Separations

12.1 INTRODUCTION

Physical separation is a process that primarily depends on physical forces to accomplish the desired separations. This approach has been used in the food industry for many years to remove haze from wine and fruit juices or nectars, separate the protein of cheese whey into fractions that have different functional property, split foreign matter from whole or milled grains, and concentrate fruit juice without heat. Physical separations can be categorized into four groups: sedimentation, centrifugation, filtration, and sieving. In the first method, two immiscible liquids (or a liquid and a solid) are separated by allowing gravity to produce an equilibrium (the heavier material will fall with respect to the lighter). Because sedimentation is often a slow process, it is often accelerated by applying centrifugal forces that increase the rate of sedimentation. This is called centrifugation or centrifugal separation. Filtration is a type of separation that stops solid particles but allows the passage of liquid, and sieving is the classification of solid particles according to diameter. This laboratory exercise will focus on centrifugal separation and sieving analyses.

12.1.1 Centrifugal Separations

Centrifugal separations have been extensively used in the food industry because separation by simple sedimentation of immiscible liquids (or of a liquid and a solid) does not progress rapidly enough to accomplish efficient separation due to small differences in component specific gravity. In these types of applications, separation can be accelerated by using a centrifugal force.

More precisely, centripetal force acts on the rotating food. The centripetal force on a particle that is constrained to rotate in a circular path is given by:

$$F = mr\omega^2 \tag{12.1}$$

where

F = centripetal force acting on the particles to maintain it in the circular path
r = radius of the path
m = mass of the particle
ω = angular velocity of the particle

Rotational speed is normally expressed in revolutions per minute so that Equation (12.1) can also be written as $\omega = 2\pi\,N/60$:

$$F = mr(2\pi\,N/60)^2 = 0.011\;mrN^2 \tag{12.2}$$

where N is the rotational speed in revolutions per minute

If this is compared to the force of gravity (F_g) on the particle ($F_g = mg$), centripetal acceleration ($0.011\;rN^2$) can be seen to have replaced gravitational acceleration (g).

The rate at which the separation of material at different densities can occur is usually expressed in terms of the relative velocity between two phases. An expression for the velocity of particles in a centripetal force field is given by the following equation [1]:

$$u = \frac{D^2 N^2 r \left(\rho_p - \rho_s \right)}{164\mu} \tag{12.3}$$

where

D = diameter of the spherical particle
N = rotational speed of the centrifuge in revolutions per minute
r = radius of the path
ρ_p = density of the particles
ρ_s = density of the solvent
μ = viscosity of the solution

In the food industry it is common to separate one component of a liquid-liquid mixture when they are immiscible but finely dispersed. For instance, in the dairy industry, milk is separated by a centrifuge into skim milk and cream.

The milk is fed continuously into a machine that is generally a bowl rotating about a vertical axis, producing cream and skim milk from its respective discharges. At some point within the bowl there must be a surface of separation between the cream and skim milk. The distance between this surface and central axis is given by the following expression [1]:

$$r_n = \frac{\rho_A r_1^2 - \rho_B r_2^2}{\rho_A - \rho_B} \qquad (12.4)$$

where

ρ_A = heavy density liquid phase
ρ_B = low density liquid phase
r_1 = radius at the discharge pipe for the more dense (heavier) liquid
r_2 = radius at the discharge pipe for the less dense (lighter) liquid

Equation (12.4) is a basic expression to be utilized in the design of the separation cylinder in which the radius at the discharge pipe for heavier and lighter liquids (r_1, r_2) can be varied independently to provide optimum separation of the two phases involved. When the discharge radius for the heavier liquid is smaller, the radius of the neutral zone decreases, but when the neutral zone is nearer the central axis, the lighter component is exposed only to a relatively small centrifugal force in comparison with the heavier liquid. Therefore, the feed to a centrifuge should be as near as possible to the neutral zone so that it will cause the least amount of disturbance to the system.

12.1.2 Sieving

Sieving is a gravity-driven mechanical size separation process that is widely used in the food industry to remove fine particles from larger ones. A stack of sieves is usually used when fractions of various sizes are to be produced from a mixture of particle sizes [2].

The sieve shaker that provides a horizontal vibration in a small-amplitude, high-frequency oscillation may be used to facilitate the sieving process. When the sieves are inclined, the particles retained on the screen fall off at the lower end and are collected by a conveyor, allowing screening and particle size separation to be carried out automatically.

Standard sieve sizes have evolved from a 25-mm aperture down to about a 0.6-mm aperture [3]. The three typical standards are the International Standard Organization (ISO), the US standard sieve number, and the Tyler sieve series. Their mesh was originally characterized by the number of apertures per inch. Table 12.1 illustrates the interrelation among the above standards. In the SI

TABLE 12.1. Standard Sieves

Aperture (m x 10⁻³)	ISO nominal aperture (m x 10⁻³)	U.S. no.	Tyler no.
22.6		⅛ in.	0.833 in.
16.0	16	⅝ in.	0.624 in.
11.2	11.2	⁷⁄₁₆ in.	0.441 in.
9.0	8.00	⁵⁄₁₆ in.	2½ mesh
5.66	5.60	No. 3½	3½ mesh
4.00	4.00	5	5 mesh
2.83	2.80	7	7 mesh
2.00	2.00	10	9 mesh
1.41	1.41	14	12 mesh
1.00	1.00	18	16 mesh
0.707	0.710	25	24 mesh
0.50	0.500	35	32 mesh
0.354	0.355	45	42 mesh
0.250	0.250	60	60 mesh
0.177	0.180	80	80 mesh
0.125	0.125	120	115 mesh
0.088	0.090	170	170 mesh
0.063	0.063	230	250 mesh
0.044	0.045	325	325 mesh

system, aperture ratios are measured in millimeters, whereas the Tyler series expresses them in terms of the number of openings per inch. Because there is a sufficient number of choices among woven wire sieves, the ratio for the opening size has been kept approximately constant from one sieve to the next. A normal series progresses as $\sqrt{2}:1$, and if still closer ratios are required, intermediate sieves are available as small as $\sqrt[4]{2}:1$.

Continuously vibrating sieves used in the flour milling industry employ a sieve with increasing apertures so the fine fraction at any stage can be removed. The shaking action of the sieve provides the necessary motion to make the particles fall through and also conveys the oversize particles onto the next section.

12.2 OBJECTIVES

(1) To become familiar with the centrifugation process
(2) To become familiar with the sieving process
(3) To estimate particle size distributions from sieve analysis

12.3 MATERIALS AND METHODS

12.3.1 Materials

- laboratory type centrifuge
- sieving apparatus
- screens or sieves
- balance
- test tubes
- raw milk
- milk powder
- instant coffee

12.3.2 Procedures

12.3.2.1 *CENTRIFUGAL SEPARATION*

Centrifugal separators are available in small-diameter, high-speed machines and those of large diameter and low speed. The centrifugal force developed in the former is generally greater than the latter. Industrial machines usually develop RCF values in the range between 700 and 22,000 g, whereas laboratory "bottle" centrifuges can develop up to 34,000 g and ultracentrifuges up to 360,000 g. Some centrifuges are best suited to liquid-liquid separation, others to solid-liquid separation, and many can perform both tasks. In this laboratory practice, a laboratory bottle centrifuge is used.

(1) Locate the centrifuge, test tubes, and milk.
(2) Become familiar with the centrifuge operation by reading the operation manual.
(3) Transfer milk to 2/3 test tube capacity.
(4) Place the test tubes into the holder of the centrifuge and close the cover.
(5) Set the centrifuge to 3000 rpm and turn it on for 10 min.
(6) When the centrifuge comes to a complete stop, open the cover and take the test tubes out.
(7) Measure the height of cream and skim milk and record the observations.
(8) Repeat Steps 3 to 8 using 5,000, 7,000, and 10,000 rpm as centrifuge speeds.

12.3.2.1.1 *Sieving Analysis*

In obtaining meaningful sieve analysis data, three major steps must take place: (1) sampling, (2) the actual sieving technique, and (3) computation, presentation, and data analysis.

(1) Locate a sonic sifter or equivalent sieve analyzer, electronic balance, and milk powder.

(2) Use the coning and quartering technique for sampling.

(3) Become familiar with the sonic sifter operation by reading the operation manual.

(4) Make sure to choose the screens in the correct sequence (in the case of milk powder, the following is recommended: 30, 40, 50, 70, 100, and 120).

(5) Weigh a sample of the material to be tested and introduce it to the top of the completed sieve stack, shake for 5 min, and then weigh the residue in the pan and calculate the percentage in relation to the starting weight. Reassemble the stack and shake for one additional minute. Weigh the residue again and calculate its percentage, and if it has increased more than 1% in 5 to 6 min, reassemble the stack and shake for another minute. The data can be plotted as a percentage throughput versus time for each data point calculated. When the change in the percentage of fines passing in the 1-min period drops below 1%, the test can be considered complete. Record the total testing time for subsequent analysis.

(6) Once the sieving interval is complete, the residue on each sieve should be removed by pouring the residue into a suitable weighing vessel. To remove material wedged in the sieving openings, the sieve must be inverted over a sheet of paper or other suitable collector and the underside of the wire cloth brushed gently with a 1-in brittle nylon paint brush. (The side of the sieve frame may be tapped gently with the handle of the brush to dislodge the particles between brush strokes, but at no time should a needle or other sharp object be used to remove particles lodged in the wire cloth. Special care should be taken when brushing sieves finer than 80 mesh because brushing can cause distortions and irregularities.) Repeat the procedure for each sieve in the stack and contents of the pan.

(7) The weight retained on the individual sieves should be added and compared with the starting sample weight. Wide variations or sample losses should be determined immediately. Compare the weights on each sieve with the total sample weight and calculate the percentage of the total retained at that point.

12.4 EXPECTED RESULTS

(1) *Centrifugal separation:* For a given material (e.g., raw milk), discuss the effect of RCF on the completeness of separation.

(2) *Sieving analysis:* Present and analyze the resulting data by plotting the graph with the following format: (1) the cumulative percentage of material retained on a screen (on a logarithmic scale) versus screen open size

(on a linear scale). The resulting curve allows a quick approximation of the sieve size at the 50 percentile point of accumulation. (2) the size distribution curve that allows smoothness of distribution by revealing the presence of sample bimodal blends.

12.5 EXAMPLES

12.5.1 Example 1

If a centripetal force of 5000 g in a small centrifuge with an effective working radius of 10 cm is established, at what speed would the centrifuge have to rotate? If the actual centrifuge bowl has a minimum radius of 9 cm and a maximum of 10 cm, what is the difference in the centripetal force between the minimum and maximum radii?

12.5.1.1 SOLUTION

(1) The centripetal force can be calculated from Equation (12.2):

$$F_c = 0.01 \; mrN^2$$

The gravity force is:

$$F_g = mg$$

therefore:

$$F_c/F_g = 0.011 \; mrN^2/mg$$

$$= 0.011 \; rN^2/g$$

Because it is already known that:

$$F_c/F_g = 5000$$

then:

$$0.011(0.1) \, (N)^2/9.81 = 5000$$

Solve for N:

$$N = 6678 \; \text{rpm}$$

(2) The minimum force is:

$$F_c/F_g \text{ (min)} = 0.011 \; r_{min} N^2/g$$

and the maximum force is:

$$F_c/F_g \text{ (max)} = 0.011 \; r_{max} N^2/g$$

Therefore, the difference is:

$$F_c/F_g \text{ (max)} - F_c/F_g \text{ (min)}$$

$$= 0.011 \; r_{max} N^2/g - 0.11 \; r_{min} N^2/g$$

$$= 0.011 * N^2 (r_{max} - r_{min})/g$$

$$= 0.011 * 66{,}782(0.1 - 0.09)/9.81$$

$$= 500 \text{ g}$$

12.5.2 Example 2

Consider that the solid particles in a liquid-solid suspension are centrifugally separated by a centripetal force with an average particle diameter of 180 μm and density of 820 kg/m³. If the liquid is water with a density of 993 kg/m³, the effective radius for separation is 10 cm, and the required velocity for separation is 20 mm/sec, determine the required rotation speed for the centrifuge.

12.5.2.1 SOLUTION

Use Equation (12.3) with a slight modification:

$$N^2 = \frac{164 \mu u}{D^2 r \left(\rho_p - \rho_s\right)} \tag{12.5}$$

where

$D = 180 \; \mu m = 1.8 \times 10^{-4} \text{ m}$
$r = 10 \text{ cm} = 0.1 \text{ m}$
$\mu = 1 \text{ cP} = 1.0 \times 10^{-3} \text{ Pa s} = 1.0 \times 10^{-3} \text{ kg/msec} = 6.0 \times 10^{-2} \text{ kg/m min}$
$u = 20 \text{ mm/sec} = 1.2 \text{ m/min}$

$\rho_A = 993$ kg/m^3
$\rho_B = 820$ kg/m^3

Substitute the above parameters into Equation (12.5):

$$N^2 = \frac{1640(6.0 \times 10^{-2})(1.2)}{(1.8 \times 10^{-4})(0.1)(993 - 820)} = 2.1 \times 10^8$$

$$N = 14,514 \text{ rpm}$$

12.5.3 Example 3

Given the following sieve analysis:

Sieve Size (mm)	% Retained
1.00	0
0.50	15
0.25	41
0.125	27
0.063	9
through 0.063	8

Plot a cumulative size curve and size distribution curve and estimate the weights (per 100 kg of powder) that would lie in the size ranges between 0.15, 0.20, 0.25 and 0.35 mm.

12.5.3.1 SOLUTION

The cumulative percentages can be calculated from the original data:

Less than aperture (mm)	Percentage (cumulative)
1.00	100
0.50	85
0.25	44
0.125	17
0.063	8

Both size distribution curve and the cumulative size curve are plotted on Figure 12.1.

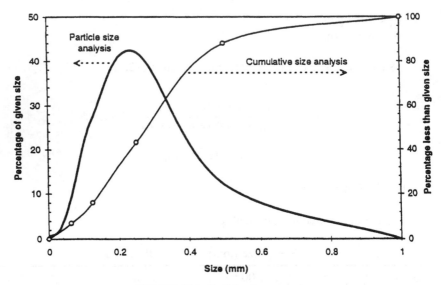

FIGURE 12.1. Particle size analysis.

The specified sizes, which can be directly read from the cumulative size analysis curve in Figure 12.1, are the following:

- The percentage of particles with size smaller than 0.15 mm is 23%.
- The percentage of particles with size smaller than 0.20 mm is 33%.
- The percentage of particles with size smaller than 0.25 mm is 44%.
- The percentage of particles with size smaller than 0.15 mm is 68%.

Therefore, the fraction of particle between 0.15 and 0.20 mm is:

$$33\% - 23\% = 10\%$$

The fraction of particle between 0.25 and 0.35 mm is:

$$68\% - 44\% = 24\%$$

Sieve analysis for particle size determination should be treated with some caution especially for particles deviating radically from spherical shape and needs to be supplemented with a microscopical examination of the powders. The size distribution of powders can be useful to estimate parameters of technological importance such as the ease of dispersion in water of a dried milk powder or the performance characteristics of a separating cyclone in a spray dryer.

12.6 REFERENCES

1. Heldman, D. R. and R. P. Singh. 1981. *Food Process Engineering.* 2 ed. AVI, Westport, CT.
2. Toledo, R. T. 1991. *Fundamentals of Food Process Engineering.* 2nd ed. AVI, Westport, CT.
3. Earle. R. L. 1983. *Unit Operations in Food Processing.* 2nd ed. Pergamon Press, New York.

abstract, 9
anemometer, 70-71, 81
apparatus, 9
atomization, 79, 86, 92

B, 43
balance
    energy, 3-4, 6, 25, 86-87, 118-119, 121
    heat, 83, 115
    mass, 1-2, 86, 115, 118-119, 121
    moisture, 82
Bacillus stearothermophillus, 50, 52
barometric leg, 119
basic principles
        (of food process engineering), 1
Biot number, 23-24, 26-27
blanching, 23
    steam, 24
bleeders, 31
blowers, 79-80
boiling point, 114-117, 120-121

calculations, 9
centrifugation, 125, 128-130, 132
centripetal
    acceleration, 126
    force, 126, 131-132
Clausius-Clapeyron relationship, 99,
        101-102, 113
Clostridium botulinum, 29, 42, 52
Clostridium sporogenes, 48
cold point, 32, 37, 39-40, 43
come-up time, 29
computer board, 25
concentration, 2, 13, 111-112
conclusions, 10
condensate, 94-95, 99, 103, 112-113, 117,
        119-120, 122
conduction, 33, 39-41, 52, 96
consistency index, 19

constant rate, 64, 68-69, 74, 78, 85, 89-90
container geometry, 40
convection, 33, 39-41, 52, 85
cooking, 29
cooling
    water, 3, 45
    cycle, 30
    lag factor, 48
    tower, 119
Coxiella burnetti, 50
cumulative
    percentage, 130, 135
    size curve, 133
    size analysis, 134

D value, 45, 49
data
    acquisition, 70
    analysis, 10
    logging system, 32, 58, 64
dehydration, 64, 111
density, 17, 19, 25, 90, 106, 126-127, 132
die, 106, 108-109
differential scanning calorimeter, 54
diffusion, 85
dilute solutions, 54
droplets, 79, 85-86, 90
drying, 23-24, 63, 72, 79, 89, 93
    air, 63-64, 84, 89
    cabinet, 78
    chamber, 64, 79-80, 83-87, 91, 95, 102
        dimensions, 91
        pressure, 95
    contact, 63
    curve, 65, 67, 69, 71-72, 74-75, 77, 102
    flux, 69
    rate, 64, 66-69, 72, 75-76, 85, 95, 102
    solid, 86
    spray, 63, 79-81, 83-86, 92, 111, 134